AIR POLLUTION AND CLIMATE CHANGE:
THE BIOLOGICAL IMPACT

SECOND EDITION

AIR POLLUTION AND CLIMATE CHANGE:
The Biological Impact

SECOND EDITION

ALAN WELLBURN

Longman
Scientific &
Technical

Copublished in the United States with
John Wiley & Sons, Inc., New York

Longman Scientific & Technical
Longman Group UK Limited
Longman House, Burnt Mill, Harlow
Essex CM20 2JE, England
and Associated Companies throughout the world

Copublished in the United States with
John Wiley & Sons, Inc., 605 Third Avenue, New York NY 10158

© Alan R. Wellburn 1994

First published 1988
Second edition 1994

British Library Cataloguing in Publication Data
A catalogue entry for this title is available from the British Library.

ISBN 0-582-09285-X

Library of Congress Cataloging-in-Publication data

Wellburn, Alan, 1940–
    Air pollution and climate change : the biological impact/Alan
Wellburn. – 2nd ed.
        p.   cm.
    Rev. ed. of: Air pollution and acid rain. 1st ed. 1988.
    Includes bibliographical references and index.
    ISBN 0-470-20006-5
    1. Air – Pollution – Environmental aspects.   2. Climatic changes –
Environmental aspects.   3. Bioclimatology.   I. Title.
QH545.A3W45   1994
574.5'222 – dc20   93–31701
                                                                              CIP

Set by 15 in 10/12 pt Times
Produced by Longman Singapore Publishers (Pte) Ltd.
Printed in Singapore

CONTENTS

PREFACE

'Behold, how great a matter a little fire kindleth.'
Epistle of St James, Chapter 3, Verse 5.

Since writing the first edition of this book, entitled *Air Pollution and Acid Rain: The Biological Impact*, many changes in our perception of problems associated with air pollution have occurred. New threats have been identified, especially in aspects of climate change associated with global warming and enhanced fluxes of ultraviolet radiation with stratospheric ozone depletion. Many of the other problems, however, still remain and most have worsened.

The opportunity to write a new edition has allowed me to update what still applies from the first edition in more concise terms and to insert fresh material on aspects of climate change. The previous three-part emphasis has been retained whereby an outline of the physical events is usually followed by a detailed consideration of the impact of pollution on both vegetation and humans in each chapter. The emphasis on explaining events inside living organisms, which are so often treated as 'black boxes' elsewhere, has been retained but the biochemical details have been reduced and simplified as much as possible. To save space, many abbreviations have had to be used throughout. A glossary is given in Appendix 3 for those unfamiliar with some of them.

Several new chapters have appeared, some by subdivision and revision while others are entirely new. In addition to global warming (Chapter 8) and stratospheric ozone depletion (Chapter 7), Chapter 4 deals with reduced air pollutants such as ammonia and sulphides, and Chapter 9 (Other Pollutants) has been extended to include lead, radon, and formaldehyde.

The publishers have also allowed me to insert more illustrations, including some in colour. I could have used this particular allowance to show plumes of smoke, cyclists wearing face masks, horrific photographs of skin cancers caused by ultraviolet light, etc. However, I have chosen to use one plate for satellite maps, etc., and the other three for effects on plants. I hope I might be excused in this, but colour is the only means of demonstrating several points mentioned in the text. Moreover, it is very difficult to find typical pictures of air pollution injury on plants anywhere in print that is easily accessed. I hope they will be useful and take this opportunity of thanking all those who sent me pictures or gave me permission to use their work. I also thank my close colleague, Deborah Robinson, who has redrawn most of the original line drawings as computer graphic images and has translated many new diagrams in a similar way.

ALAN WELLBURN
April 1993

ACKNOWLEDGEMENTS

We are grateful to the following for permission to reproduce copyright material:

Der Bundesminister für Forschung und Technologie for Fig. 1.1 (from *Initial Report on Research on Environmental Damage to Forests*, CCRAC 84–17, Annex 1, 1984); the authors, I. Nicholson *et al.*, for Fig. 1.4 (from *Ecological Impact of Acid Precipitation*, SNSF Project, Oslo, 1984); the late Prof. E. B. Ford for Fig. 1.9; D. Reidel Co and the authors for Figs. 2.1 and 3.5 (R. A. Cox and S. A. Penkett in *Acid Deposition*, Beilke & Elshout (eds) 1983) and Fig. 10.1 (H. Fluhler in *Effects of Accumulation of Air Pollutants in Forest Ecosystems*, Ulrich & Pankrath (eds) 1983); The Watt Committee on Energy Ltd and the author, F. B. Smith, for Fig. 5.1 (*the Watt Committee Report No. 14*, 1984); the WHO Centre on Surface and Ground Water, Canada and the authors, D. Whelpdale after G. Gravenhorst, for Fig. 5.2, from the *Water Quality Bulletin*, April 1983; the authors, D. Sutcliffe and T. Carrick, for Fig. 5.5 (Effects of acid rain on waterbodies in Cumbria, in *Pollution in Cumbria*, ITE Symposium, 1985); CRC Press Inc and the authors for Fig. 11.1 (P. H. Freer-Smith and A. Wellburn in *Models in Plant Physiology/Biochemistry*, Newman & Wilson (eds) 1986); M. Whitmore and T. A. Mansfield for Plate 1; C. Hufton for Plate 2; A. Posthumus for Plates 3, 5 and 6; M. Treshow for Plate 4; NASA for Plates 7 and 9; K. Bull for Plate 8; F. Bauer for Plate 10; N. Paul for Plates 11, 12 and 13.

Whilst every effort has been made to trace the owners of copyright material, in a few cases this has proved impossible and we take this opportunity to offer our apologies to any copyright holders whose rights we may have unwittingly infringed.

To Florence

Front cover: Ponderosa pines (*Pinus ponderosa*) in the San Bernardino mountains in southern California showing severe injury due to photochemical smog (*Courtesy of Dr Mark Poth, US Forest Service, Riverside, USA*)

Back cover (*inset*)*:* Fluoride injury to Ponderosa pine (*Pinus ponderosa*) (*Courtesy of Prof. Mike Treshow, Salt Lake City, USA*)

Chapter 1

INTRODUCTION

'We have first raised a dust and then complain we cannot see.'
from 'The Principles of Human Knowledge' by Bishop Berkeley (1685–1753).

Definitions and terms

A chemical that is in the wrong place at the wrong concentration is either a pollutant or a contaminant. The distinction resides in the ability of a pollutant to injure humans or parts of the biosphere upon which they depend. The study of air pollution covers those pollutants that are emitted into the atmosphere (usually from the land) as gases or particulates which then, directly or indirectly, degrade or adversely affect physical and biological systems.

In the atmosphere, pollutants may move from a dry, gaseous phase into a liquid phase before falling to the surface. If acidic in nature this is popularly known as acid rain. A less emotive and more descriptive word to cover acidic and non-acidic precipitation, both as droplets and as snowflakes, etc., is wet deposition. Wet and dry deposition are the two main pathways by which atmospheric pollutants are returned to the surface of the Earth. Other terms like rain-out and run-off are also used (see Fig. 1.1).

Atmospheric emissions of pollutants are caused by human activity. However, now, and long before the appearance of humans, natural emissions of similar substances by volcanoes and swamps have caused considerable disruption to the local environment. Consequently, natural emissions must be taken into account when considering air pollutants and their sources. The term anthropogenic is sometimes used to separate human-induced from natural emissions. However, as anthropogenesis means the study of the origin and development of humankind, the use of the word anthropogenic in the context of pollution is clearly wrong. Man-made is sexist and human-induced is ungainly so perhaps homogenic is better.

Damage by air pollutants to metals, fabrics and materials used by humans is very evident but the biological effects of air pollutants directly upon humans or upon the living systems that surround them are often subtle and far more significant. Moreover, biological effects are detected

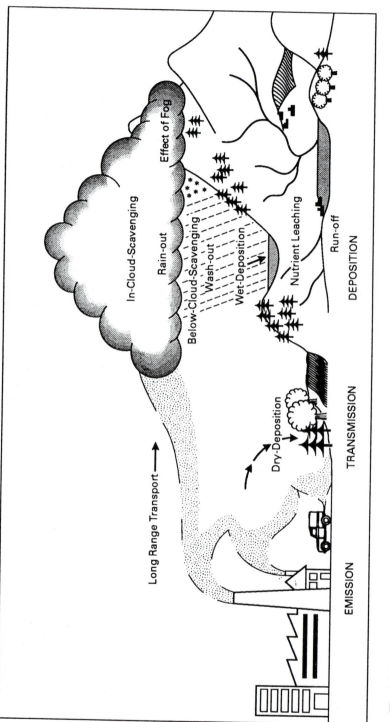

Fig. 1.1. Emissions and depositions of atmospheric pollutants (courtesy of the Bundesminister für Forschung und Technologie, Germany).

Table 1.1 Different types of suspended particulates in the atmosphere

Name[a]	Nature	Diameter
Grit	Solids, settle out quickly	$> 500\ \mu m$
Dust	Solids, setting more slowly	$2–500\ \mu m$
Smoke	Gas-borne solids	$< 2\ \mu m$
Mist	Liquid droplets	$0.1–2\ \mu m$
Aitken nuclei	Solid or liquid droplets	$0.003–0.1\ \mu m$

[a] Aerosols appear to have a number of different interpretations. They may consist of solid or liquid droplets which are either less than $1\ \mu m$ in diameter or include mechanically generated particles greater than $1\ \mu m$ which remain suspended for long periods. If they collectively obscure visibility, they then form a haze.

faster and at lower concentrations than those upon non-living materials. Although damage to inanimate materials has a huge economic cost, it has always been accepted that danger to human health was the main reason for air pollution controls. More recently, especially in the case of air pollution, controls have also been imposed to protect vegetation and biological ecosystems upon which humans depend. This is because, in the case of many air pollutants such as sulphur dioxide (SO_2), nitrogen dioxide (NO_2), and ozone (O_3), the plants that make up major parts of these ecosystems are more sensitive than humans to air pollutants. Consequently, in each of the early chapters that follow, pathways of major pollutants are considered from the point of emission into the atmosphere and from there into plants and animals. Once inside each, the corresponding pollution effects are then examined in detail.

Air pollutants exist in particulate and gaseous forms. Particulates are particularly diverse in character (see below) but gaseous air pollutants may be separated into primary and secondary forms. Primary air pollutants like SO_2, most nitrogen oxides, carbon monoxide (CO), and unburnt hydrocarbons (HCs) are emitted directly into the atmosphere. Secondary air pollutants, such as O_3 and peroxyalkyl nitrites (PANs), are formed later by atmospheric reactions involving primary air pollutants and strong light.

Particles and aerosols

Air-suspended matter consists of various solid or liquid droplets which have a number of different names. Each of them has a specific meaning despite being popularly used in an interchangeable manner, and Table 1.1 distinguishes them by nature and size. In the case of grits and most coarse dusts, only particles greater than $50\ \mu m$ in diameter are visible and those below $10\ \mu m$ quickly settle close to the source of pollution. Those less than $10\ \mu m$ across remain suspended for some time and, if very small (0.1–2 μm), act as nuclei for the condensation of water during cloud formation. They are only removed as rain or when hit by falling rain (wash-out). If they escape from the lower troposphere to upper altitudes, however, they remain there for months or even years. Extremely small particles (0.005–0.1 μm) are known as Aitken nuclei and are mainly formed as condensation products from hot vapours. Aitken nuclei eventually coagulate to form larger particles (0.1–2 μm) which are gradually removed by rain but unburnt HCs also assist with the formation of these larger nuclei.

Particulates contain a wide diversity of water-soluble and insoluble components. The latter usually consist of elemental carbon, iron oxides, and a variety of other materials such as rock-derived powders (e.g. quartz, mica, china clay, and fibres such as asbestos) – all of which are known to cause dust-related lung diseases such as pneumoconiosis, silicosis, and asbestosis. Toxic metals like lead or cadmium also contribute to the insoluble components of particulates and have their own toxicological problems (see Chapter 9).

Soluble components of particulates include the common cations (Na^+, K^+, Ca^{2+}, Mg^{2+}, NH_4^+, and H^+) as well as chloride, sulphate and nitrate. Sodium, chloride and magnesium arise naturally from sea-spray while calcium and potassium are usually derived from soil-blown material. Ammonium, sulphate, nitrate and protons (H^+), however, are produced by both volcanic and homogenic emissions.

Reductions in visibility due to aerosol hazes are caused by the combination of ammonia with the oxidation products of atmospheric pollutants forming reflective particles of NH_4HSO_4, $(NH_4)_2SO_4$ and NH_4NO_3. Such particulates also absorb and reflect incoming solar radiation. In periods before smoke control legislation as much as 50% of incoming light to cities like London was lost in the winter (10% in summer). Even now, with ammonium-containing hazes, there may be as much as an average 15% yearly loss of sunlight as well as an increased likelihood (up to 10%) of cloud cover because of the nucleating effect of small particulates on cloud formation.

If they fall on buildings, larger particulates accelerate the atmospheric pollution-induced corrosion of stonework, etc. However, if they adhere to leaf surfaces, they reduce photosynthesis by blocking stomatal pores or by reflecting and absorbing incident light. The great diversity of size and composition makes particulates difficult to monitor. Most measurements rely upon weighing total suspended particulates collected by filtration or by determining the staining capacity of air upon filters.

Standards vary from country to country but, for example, the particulate air quality standard in the USA is 75 mg m^{-3} a^{-1} or 260 μg m^{-3} as a daily (24 h) average not to be exceeded for more than one day per year. Enforceable limit values of smoke set by the Commission of the European Communities (CEC) are 80 μg m^{-3} (as a median of daily mean), 130 μg m^{-3} (as a median over the winter months October to

Table 1.2 Annual particulate depositions (t km^{-2}) between 1916 and 1922 in UK towns (after Ashworth, 1933)

Year (*April to March*)	*London*	*Glasgow*	*Newcastle upon Tyne*	*Rochdale*	*Malvern*
1916–17[a]	157	149	279	292	35
1917–18[a]	149	175	249	417	31
1920–21	120	126	199	298	30
1921–22	111	100	211	203	27

[a] The wartime increase in extra pollution is equivalent to what might be expected if $1\frac{1}{2}$ extra days were being worked. With overtime, this was the case.

March) and 250 µg m^{-3} (as the 98th percentile of daily values). Guide values (to which CEC members should aim) are 40–60 µg m^{-3} smoke as an arithmetic mean of daily values or 100–150 µg m^{-3} as a daily mean.

These and similar regulations and recommendations have done much towards cleaning the atmospheres above developed countries in terms of visibility and amenity. It is difficult for the young in developed countries to appreciate just how bad particulate pollution was in the recent past – and still is in developing countries. Table 1.2, which shows estimates of particulate deposition in certain areas of the UK during 1916–22, demonstrates the severity of the problem. Rochdale was once one of the most important cotton-spinning towns in the world (no longer) and Malvern a health resort to the south-west (and cleaner) side of Birmingham. From figures available now, present depositions for Rochdale are less than those for Malvern then – a clear measure of how much improvement can be made through legislative measures.

Gaseous units

Concentrations of gaseous air pollutants are expressed either as mass per unit volume (µg m^{-3}) or as volume per unit volume. In the non-SI past, the latter was usually expressed as parts per million (ppm) or parts per billion (ppb) but in SI terms (see Appendix 3) these units are µl l^{-1} and nl l^{-1}, respectively. Unfortunately, neither µg m^{-3} nor µl l^{-1} is entirely satisfactory. It might be better if mass per unit mass (i.e. µg g^{-1}) were to be used, which would then bring air concentrations into line with conventional SI expressions of concentration in solids or liquids. Unfortunately, it is difficult to conceive of what volume of air corresponds to 1 g and this varies with altitude anyway. Furthermore, any mass per unit volume measurement must take into account changing temperatures as well as pressure, whereas the volume per unit volume of an ideal gas is independent of temperature or pressure.

Most air pollutants behave for all practical purposes as ideal gases. Consequently, the relationships between the two sets of units are given by the following expressions:

$$\mu g\ m^{-3} = \frac{nl\ l^{-1} \times M \times 10^{-3} \times T_0 \times P}{V_0 \times T \times P_0}$$

$$nl\ l^{-1} = \frac{\mu g\ m^{-3} \times V_0 \times 10^{-3} \times T \times P_0}{M \times T_0 \times P}$$

where M is the molecular weight and V_0 is the molar volume of an ideal gas which is equivalent to $22.4 \times 10^{-3}\ m^3\ mole^{-1}$ when standard temperature (T_0 in Kelvin) and pressure ($P_0 = 101.3$ kPa) prevail.

The important point is that mass per unit volume expressions do not allow an immediate comparison of one pollutant with another in terms of numbers of molecules. By contrast, an atmosphere containing 1 μl l^{-1} SO_2 contains the same number of pollutant molecules as one containing 1 μl l^{-1} O_3 or NO_2 because the molecular weight and the volume of gas containing 1 gram molecule have already been taken into account. For this reason, this book uses μl l^{-1} or nl l^{-1} as far as possible but approximate interconversions for common air pollutants are provided in Appendix 3.

In some cases, however, μl l^{-1} or nl l^{-1} is inappropriate – as in the case of particulates or aerosols. Mass per unit volume measurements must then be qualified by the particle sizes or the droplet diameters.

Atmospheric composition and climate

'A solid has volume and shape while a liquid has volume but no shape and a gas has neither volume nor shape', which means that gas or vapour is matter in a perfectly fluid state. The whole planetary atmosphere behaves similarly and is about 1000 km high at the Equator and rather less (800 km) at the Poles. The total mass of the atmosphere is 5.14 Pt (where P = peta = 10^{15}; 1 metric tonne (t) = 1000 kg) but more than 99.9% of that mass of gas is below an altitude of 50 km.

The atmosphere consists of a mixture of gases, the major ones being N_2 (78.08%), O_2 (20.95%), argon (0.93%), CO_2 (0.035%), neon (0.0018%), helium (0.005%), and krypton (0.0001%). However, concentrations of atmospheric gases, water vapour and air pollutants are not uniform with altitude. Figure 1.2 shows some of their typical concentrations (or mixing ratios) upwards through the atmosphere.

There are distinct warm and cold layers above the surface of the Earth (Fig. 1.3). Closest to the surface is the troposphere where there is a decrease of temperatures with height up to 12 km. Above this is the stratosphere (12–50 km) where temperatures increase with height.

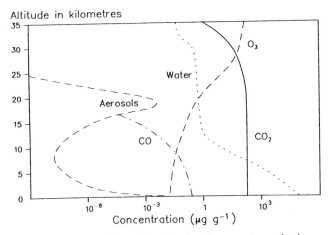

Fig. 1.2. Concentration profiles with altitude of some atmospheric components showing the rise of O_3 and aerosols upwards into the stratosphere and the steady fall of H_2O and CO_2 with increasing height.

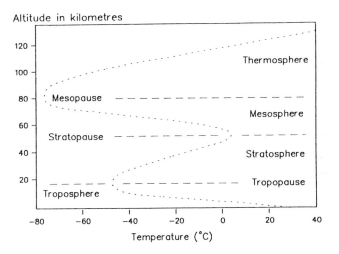

Fig. 1.3. Temperature gradients through the different regions of the atmosphere.

Uppermost are the mesosphere (50–80 km), where temperatures decrease again, and the thermosphere (80 km upwards) where they rise once more, sharply so after 110 km. The boundaries where inversions of temperature take place are called the tropopause, the stratopause and the mesopause respectively (Fig. 1.3). Temperature changes in the atmosphere are brought about by the absorption and partial re-emission of visible, ultraviolet and near infrared solar radiation by certain atmospheric gases, especially carbon dioxide (CO_2), methane (CH_4), O_3 and water vapour – the greenhouse gases (see Chapter 8).

The atmosphere is also warmed by combustion processes on land as well as by the absorption of reflected thermal radiation. Together, these may add an extra 20% to that caused by incoming solar radiation. As the temperature of air masses near the surface rises, their density falls, causing them to expand and rise. Heat is then lost to the higher but cooler regions of the troposphere, causing these air masses to fall once again. This gives an up-and-down movement to tropospheric air. Vertical velocities induced in this way may reach 20 m s^{-1} in thunderstorms but 10 cm s^{-1} is more normal.

Polluted air masses are also moved sideways by winds that mainly blow from west to east in the middle latitudes of both hemispheres. At 30°N, for example, a 35 m s^{-1} west to east flow takes about 12 days to get round the world. North–south oscillations also occur at the same time as the west to east flows which, when combined with up and down movements, impose a rotary turbulence to air masses at low altitudes which is called Hadley-cell circulation. This is a dominant feature at the Equator but at middle latitudes (where most atmospheric pollution is emitted) north to south eddies in west to east prevailing winds are the dominant features.

Exchanges between the troposphere and the stratosphere take place mainly at the Equator as warm air masses rise over the tropics but are prevented from rising beyond that because temperatures rise with altitude above the tropopause. This means that traces of pollutants remain in the equatorial stratosphere much longer (often years) and often move either north or south before mixing and moving into the troposphere. This mechanism is the major mechanism by which atmospheric pollutants are transferred between hemispheres.

The troposphere contains varying amounts of water vapour (Fig. 1.2) – the warmer the air, the more water vapour is retained. But when saturated and then cooled, water condenses to form clouds, mists or fogs. Mists differ from fogs by virtue of having smaller droplet sizes (< 2 μm). Water vapour produced by combustion is an insignificant component of the global water cycle, but fogs are often associated with urban and industrial areas because the particulate pollution provides the necessary nuclei for the condensation of water, especially at night when temperatures are lower. By contrast, the lower stratosphere holds only minute amounts of moisture because, as it enters the tropical tropopause, it is frozen out at temperatures approaching -80°C to form high cirrus cloud.

Annual global precipitation rates over land are about 71 cm a^{-1} while evaporation rates are much lower (47 cm a^{-1}). Over oceans, which are more than double the land surface area, precipitation (110 cm a^{-1}) and evaporation (120 cm a^{-1}) rates are higher. This means there is a net

Table 1.3   Approximate residence times for atmospheric gases in the atmosphere

Component	Time (years)
NH_3, H_2S	$< 0.005^a$
NO, NO_2	$< 0.01^a$
SO_2	$< 0.02^a$
H_2O	0.03
O_3	0.3
CO	0.4
CH_4	3
N_2O	7
CO_2	10
O_2	8000

^{a} Highly variable, depending upon moisture content and other factors.

transfer of 0.04 Pt of water from the oceans to the land. This is compensated by an equivalent amount of run-off from the land to the sea. However, because the atmosphere is such a small water reservoir by comparison to the land or the oceans, water transfer through the atmosphere from sea to land is relatively quick.

The residence time of a substance (e.g. SO_2) in a reservoir is equivalent to the total capacity of SO_2 divided by the input or the outflow of SO_2. In the case of atmospheric water, total capacity (0.0013 Pt) divided by net flow from oceans to land (0.045 Pt a^{-1}) gives 0.03 years or < 11 days as the residence time. By comparison to the residence times from other atmospheric gases (Table 1.3), this is a relatively short time. Only gases like CO_2, which have residence times longer than half a year, have sufficient mixing to produce relatively uniform global concentrations.

Deposition rates

Air pollutants are removed from the atmosphere by a variety of different processes (Fig. 1.1) and, individually, at very different rates. This is because some are readily absorbed or adsorbed by certain surfaces while others are relatively unreactive. Indeed, there is a wide variation in absorption and adsorption of gases like NO, NO_2, SO_2, and O_3 by various surfaces (Table 1.4). These uptake factors then determine how far and for how long a pollutant travels from its source before returning to surfaces. Over water, for example, atmospheric NO will travel much further and for longer than NO_2 but over soils there is very little difference between them. Similarly, SO_2 will travel hundreds of kilometres over snow-covered soil which is a poor sink but only a few kilometres over moist, unfrozen soil.

Table 1.4   Deposition velocities (mm s^{-1}) of atmospheric pollutants towards different surfaces

Surface	Pollutant				
	NO	*NO$_2$*	*PANs^a*	*SO$_2$*	*O$_3$*
Soils	1.9	1.6	2–30	2–11	2.5–10
Seawater	0.015	0.15	0.2	2	0.5
Freshwater	0.007	0.1	—	1	0.1
Plants	1	4–60	6	1–29	1–17

^{a} See Chapter 6.

The gases listed in Table 1.4, with the partial exception of NO, all show high rates of deposition towards plant surfaces. This is because for NO$_2$, O$_3$, etc., physiological processes such as the resistance to gaseous flow offered by the stomatal apertures have a dramatic effect on deposition rates. When open, the stomata give access to much larger absorptive areas inside leaves and hence the rates of deposition are higher. The remaining barrier for uptake is then the boundary layer of still air above and below leaves. However, if it is windy, this is blown away and the boundary layer resistance much reduced (see Chapter 2).

There is another group of gases which are so reactive that their uptake is not impaired by surface resistances. The deposition of gases like HNO$_3$, HCl, HF and NH$_3$ is controlled solely by atmospheric resistances. Consequently, they all have large deposition velocities (> 20 mm s^{-1}) in windy conditions (> 4 m s^{-1}) when boundary layer resistances are removed.

Levels and limits

For centuries in England, as in many states of Europe, the law has protected individuals from point sources of air pollution. In 1691, for example, the baker next door to Thomas Legg of London was made to put up a chimney 'soe high as to convey the smoake clear of the topps of the houses'. But when point sources become larger, more numerous and merged into polluted environments, atmospheres as well as individuals have to be protected by central governments using legislation. Finally, when pollution from one national state causes problems in another country, international agreement is required.

Sources of air pollution are diverse and it is often impossible to discriminate between them. Domestic sources often surround industrial sources and both require power generation. Individuals, as well as commercial enterprises, also contribute to the atmospheric problems that arise from transport or the disposal of waste. Table 1.5 therefore shows

Table 1.5   Approximate proportions (%) of generated atmospheric pollution (based on 1968 figures for the USA)

Operation	Particulates	SO₂	NO + NO₂	CO	Unburnt hydrocarbons
Power and heat	47.9	74.9	53.2	2.2	2.3
Transport	6.5	2.3	42.7	76.7	59.7
Waste disposal	5.3	0.3	2.9	9.4	5.9
Others	40.2	22.4	1.1	11.7	16.6
Solvent evaporation	—	—	—	—	15.4

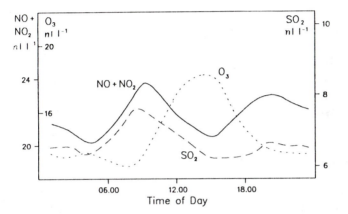

Fig. 1.4. Daily pattern of atmospheric pollutant fluctuations very typical of cloud-covered industrial regions subjected to moderate winds (courtesy of Drs Nicholson, Paterson, Cape and Kinnaird and Prof. Fowler, ITE (NERC), Scotland).

the proportions of pollution caused by various activities without distinguishing between industrial and domestic contributions.

Fossil fuel-based power generation contributes significantly to particulate, as well as SO_2 and NO_2 emissions – different fossil fuels producing various types and amounts of pollutants. Table 1.6 shows the typical ranges of pollutant gases emitted by power plants not fitted with flue gas denitrification and desulphurization.

Pollutant levels fluctuate rapidly but patterns do emerge, especially during a working day. Figure 1.4 shows a typical pattern which could be repeated in many conurbations across Europe, the eastern USA, and Japan during mild, gusty weather with partial cloud cover. Demands for power start early in the day and levels of SO_2 rise. This coincides with travel to work so amounts of nitrogen oxides (and unburnt HCs) also increase. Breaks in the cloud cover around noon may permit additional

Table 1.6   Typical ranges of gases (in $\mu l \, l^{-1}$) emerging from the stacks of power plants using different forms of fuel

Pollutant	Fossil fuel			
	Coal	*Oil*	*Gas*	*Peat*
SO_2[a]	400–3000	500–3000	—	100–1000
NO	300–1300	250–1300	200–500	100–1000
NO_2	10–40	10–40	10–30	3–30
CO	—	—	100	2000–5000
UVHCs[b]	10–60	100–2000	100–2000	30
HCl	80	—	—	—

[a] SO_3 values are in the range 1–10% of SO_2 levels.
[b] Unburnt volatile hydrocarbons.

photochemical reactions and O_3 formation which then enhance sulphate and nitrate formation. As traffic returns home, rising levels of nitrogen oxides then scavenge the O_3 away. A late pulse of SO_2 then signifies preparation for the evening meal and additional domestic heating.

Every developed country has legislation to control or limit emissions of atmospheric pollution and it is not within the remit of this book to consider individual examples of how they operate. However, international agreements do exist to ensure desirable air quality standards. The Commission of the European Communities (CEC), for example, uses both limit values and guide values. The difference between the two is that the former have to be enforced by national legislation while the latter have no legal implication but indicate that individual CEC governments should work towards achieving these levels. Superimposed on these are air quality guidelines recommended by the World Health Organization (WHO) which are to be regarded as desirable objectives and, in the main, are lower than CEC guide values. Table 1.7 summarizes these three types of international recommendations.

In the workplace, standards based upon permissible exposure limits or threshold limit values (TLVs; see also Appendix 3) are set out in the USA for specific substances, including air pollutants, to which it is believed the majority of working adults may be exposed for five 8 h working periods per week without showing ill effects. Member countries of the CEC and other developed countries usually abide by these TLVs which were originally established on the basis of a wide range of animal experiments, medical knowledge, epidemiology and other environmental studies. In the UK, for example, TLVs are known as occupational exposure limits (OELs). A number of important air pollutant TLVs are shown in Table 1.8.

It will be quickly appreciated from Table 1.8 that the TLV for SO_2, for example, is well in excess of those recommended as international air

Table 1.7 Examples of international air quality standards for SO_2 issued by the Commission of the European Communities (CEC) and the World Health Organization (WHO)

	SO_2	
Directives and Recommendations	$\mu g\ m^{-3}$	$nl\ l^{-1}$
CEC Limit Values (1980)		
Year (median of daily values)	120^a	45
Winter (median of daily values)	180^b	68
(October–M arch)		
CEC Guide Values (1980)		
Year (daily arithmetic mean)	40–60	15–23
24 hours (daily mean)	100–150	38–56
WHO Guidelines (1985)		
Year (arithmetic mean)	30^c	11

[a] $80\ \mu g\ m^{-3}$ if smoke $> 40\ \mu g\ m^{-3}$.
[b] $130\ \mu g\ m^{-3}$ if smoke $> 60\ \mu g\ m^{-3}$.
[c] The equivalent figures for NO_2 and O_3 are 30 and 60 $\mu g\ m^{-3}$ respectively.

quality standards (Table 1.7). This is because TLVs were only designed to be applicable to healthy workers over relatively short periods during their working life. On the other hand, air quality standards have to encompass long-term effects upon the total population, including the young, the old, and the ill, as well as upon vegetation, materials, and the environment as a whole.

However, irrespective of which air quality standards are set they do not of themselves create cleaner air unless policies are adopted to ensure that they are achieved. Implementation of monitoring, inspections and enforcement must be rigorous if success is to be ensured. The strictest standards are useless if there is no intention to ensure their adoption and use in the environment, in the home, or at the workplace.

Critical loads

Certain countries have taken air quality standards a stage further in terms of quantifying and controlling inputs of pollutants into sensitive areas. One of the commonest in use is the critical load which is defined as 'the highest load that will not lead in the long term (50 years) to harmful effects on the most sensitive ecological systems'. Once such loadings have been established, regulators then seek to achieve lower target loads which then ensure the critical loads are not exceeded. This approach has many merits, the chief of which is that it directly links environmental protection with deposition rates and emission controls. The initial difficulty, however, is to establish an accurate critical load

Table 1.8   Threshold limit values (TLVs) for five 8 h exposures (as a weighted average) per week for healthy humans (ACGIH values, see Appendix 3)

Pollutant	TLV $\mu l\ l^{-1}$	TLV $mg\ m^{-3}$
Benzene	10	30
CCl_4	5	31
CO	50	55
CO_2	5000	9800
CS_2	10	31
HCHO	1	1.5
HF	3	2
F_2	1	2
H_2S	10	15
NH_3	25	35
NO	25	30
NO_2	5	9
PANs	0.08	—
Organic lead	—	0.15
O_3	0.1	0.2
SO_2	5	13
Unburnt HCs	500	—

value for one particular pollutant which takes into account all the appropriate environmental factors. The other main problem is that critical loads cannot be applied in an additive manner for a mixed pollutant situation. Interactions between pollutants (see also Chapter 11) are very common and, consequently, individual critical loads cannot be applied so a fresh combined critical load must be established.

   Critical loads have been applied most effectively in assessing the risks of acid rain or total acidic wet deposition (see Chapter 5) to different areas. Plate 8 shows a typical computerized predictive map of total acidity expressed in critical load terms. On the basis of maps such as these, lower target loads are then set by individual countries or groups of countries. In North America, for example, a target loading of 20 kg wet SO_4^{2-} ha^{-1} a^{-1} is regarded as being protective of all but the most sensitive aquatic ecosystems.

Detection

Monitoring of air pollutions is a prerequisite of air quality control and is carried out by a wide variety of methods which have differing sensitivities and specificities. Some are continuous while others require batch samples collected over a period, some of which (e.g. HCs) have to be concentrated before analysis. Continuous methods are more adaptable to unattended

Table 1.9   Continuous methods of air pollutant measurement

Pollutant	Technique	Response time	Detection limit
SO_2	H_2O_2/conductivity	3 min	10 nl l^{-1}
	Flame photometric	25 s	0.5 nl l^{-1}
	Pulsed fluorescence	2 min	0.5 nl l^{-1}
NO	Chemiluminescence with O_3	1 s	0.5 nl l^{-1}
NO_2	Reduction (+ above)^a	1 s	0.5 nl l^{-1}
O_3	KI oxidation/ electrolysis	1 min	10 nl l^{-1}
	Chemiluminescence with CH_2CH_2	3 s	1 nl l^{-1}
	UV spectroscopy	30 s	3 nl l^{-1}
CO	Electrochemical	25 s	1 µl l^{-1}
	Non-dispersive IR	5 s	0.5 µl l^{-1}
Unburnt HCs	Flame ionization GLC	0.5 s	10 nl l^{-1}
	Non-dispersive IR	5 s	1 µl l^{-1}

^{a} May also be used for NH_3 but reduction carried out at temperatures higher than 650°C.

operation but are usually also more expensive. Moreover, periods of automatic sampling by continuous techniques must take into account the time taken for the detectors to respond.

Cheaper, non-continuous methods are standardized to fixed periods of sampling and often require manual operations during or after sampling. That is not to say that automatic continuous methods do not require attention. Several of them require supplies of other gases which have to be checked or they have moving parts (pumps) which require regular maintenance. Table 1.9 lists the main continuous techniques with minimum levels of pollutant detection and the 90% response time in each case.

Chemiluminescence is a good example of a continuous technique. It is used for the detection of both nitrogen oxides and ozone (O_3). For the nitrogen oxides, excited NO_2* is formed from nitric oxide (NO) in the presence of excess O_3 which then emits light at 1200 nm which can be detected by a photomultiplier. Meanwhile, NO_2 is converted to NO by heat (650°C) prior to measurement and can be separately determined by difference from the measurements of NO alone.

Ozone is determined in a similar way by reaction with hydrocarbons such as ethylene (C_2H_4, or ethene) to form a light-emitting free radical (see Appendix 1) which decays to formaldehyde (HCHO), emitting light of wavelength *ca.* 435 nm as this occurs. This means that light from O_3 measured in this way does not interfere with any light from the reaction of O_3 with NO.

Table 1.10   Non-continuous or semi-automatic methods of air pollutant measurement

Pollutant	Technique[a]	Collection period	Detection limit (nl l⁻¹)
SO₂	H₂O₂/acid–base titration	24 h	2
	West–Gaeke process	15 min	10
NO₂	Modified diazotization	30 min	5
O₃[b]	KI oxidation + spectrophotometry	30 min	10
PANs	GLC/electron capture	—	1
CO	Methanation/ flame ionization	—	10
Unburnt HCs	GLC/flame ionization	—[c]	1

[a] GLC = gas–liquid chromatography, KI = potassium iodide.
[b] Including total oxidants.
[c] Samples of unburnt (volatile) hydrocarbons require concentration prior to injection.

Older batch or semi-automatic techniques have been much improved with time. Table 1.10 lists a number of the more successful of these techniques, the period most often used for collecting the sample, and the lowest detection limit in each case. In developed countries, these procedures are currently being phased out but, being cheaper, they still have a valuable role in developing countries.

Bioindicators

Chemical monitoring of air pollutants can sometimes be replaced by biological indicators. Certain lichens, for example, are sensitive to atmospheric pollutants such as SO₂, hydrogen fluoride (HF), O₃ and PANs. Pollution mapping based on the distribution and abundance of lichen species is frequently undertaken but a number of precautions have to be taken to ensure reliability and relevance. For example, if epiphytic lichens are used then the sampling has to be from the same species of tree growing in similar environmental conditions in the absence of unusual nutrients, herbicides or pesticides, etc. A great deal then rests on experience if overlap between different pollutants and differences in climate (e.g. humidity) are to be avoided. Such surveys are used to best advantage when an unpolluted area is about to have an industrial plant located nearby. Surveys of the same site at different times may then show evidence of a deleterious change. The study illustrated in Fig. 1.5 is a good example of the use of lichens as bioindicators.

Higher plants, highly sensitive to one particular pollutant, are also used as bioindicators. A range of plants, each of which shows specific

Table 1.11   List of plant series suitable for use as air pollution bioindicators in Europe (after Steubing and Jäger, 1982)

Pollutant	Species and variety
SO$_2$	Alfalfa (*Medicago sativa* L. cv. Du Puits) (see Plate 3)
	Clover (*Trifolium incarnatum* L.)
	Pea (*Pisum sativum* L.)
	Buckwheat (*Fagopyrum esculentum* Moench.)
	Great plantain (*Plantago major* L.)
NO$_2$	Wild celery (*Apium graveolens* L.)
	Petunia sp.
	Ornamental tobacco (*Nicotiana glutinosa* L.)
O$_3$	Tobacco (*Nicotiana tabacum* L. cv. Bel W3) (see Plate 6)
PANs	Small nettle (*Urtica urens* L.)
	Annual meadow grass (*Poa annua* L.)
HF, fluorides	*Gladiolus gandavensis* L. cv. Snow Princess (see Plates 5 and 12)
	Tulip (*Tulipa gesneriana* L. cv. Blue Parrot)
General accumulators	Italian rye grass (*Lolium multiflorum* Lam. ssp. *italicum*)
	Cabbage (*Brassica oleracea* L. cv. Acephala)
Bark accumulators	Rose (*Rosa rugosa* Thunb.)
	Thuga orientalis L.

visible injury to one pollutant or accumulates certain pollutants in a characteristic manner, have been developed. A list of those indicator plants suitable for Europe is shown in Table 1.11. These plants are container-grown under clean-air conditions to a predetermined condition and age and then taken to various locations over a whole area. They are then brought back to a central location after a fixed period of exposure and compared with equivalent clean-air controls. From the visible differences, an assessment of air pollution conditions is then made. Whole countries like the Netherlands have been covered by such networks. The results from these biological indicators spread out over a wide area are then integrated with chemical monitoring data from a few selected sites and contoured maps are drawn (see Fig. 1.6).

Careful comparison of bioindicator plant response with equivalent clean-air controls is infinitely superior to subjective judgements of visible injury to pre-existing vegetation consisting of widely different species. The latter can only draw attention to the possibility of problems which may or may not be associated with air pollution damage. Many other factors like drought, insect attack, waterlogging, fungal infection, and adverse temperature can cause visible injury to vegetation and, consequently, wrong conclusions can easily be drawn. If bringing expensive chemical monitoring into a suspect locality is impossible, setting out sensitive plants can often provide definitive answers.

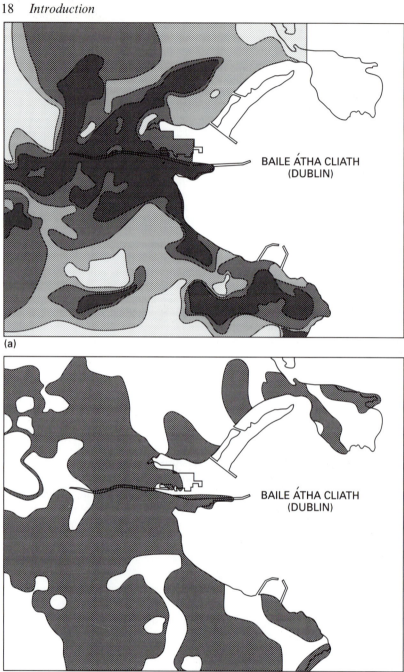

(a)

(b)

Fig. 1.5. The air pollution climate around Dublin as monitored by lichen and 'pink yeast' surveys carried out by Irish schoolchildren in 1988. (a) presumed SO_2 distribution based on a few 24 h monitors, heavy shade $\equiv$ 50 µg m^{-3} moving progressively outwards to 10 µg m^{-3}. (b) SO_2 distribution according to 'pink

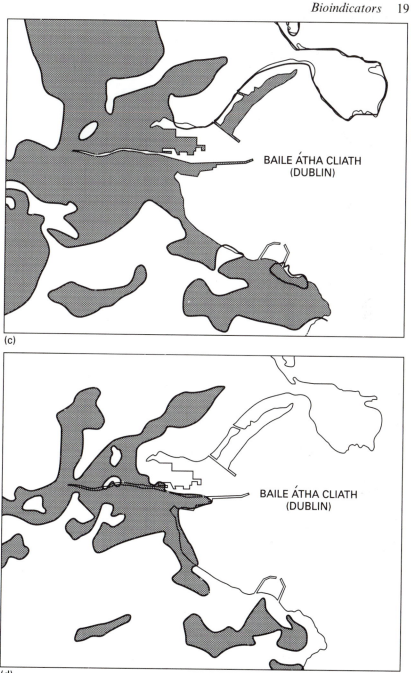

(c)

(d)

yeasts', heavy shade ≡ 'air quality very poor'. (c) distributions of the lichens *Parmelia saxatilis*, *P. sulcata* and *Hypogymnia physodes*, shading = absent. (d) distributions of the lichens *Lepraria incana* and *Lecanora conizaeoides*, shading = absent. Maps courtesy of Prof. D.H.S. Richardson and An Foras Forbartha, Eire.

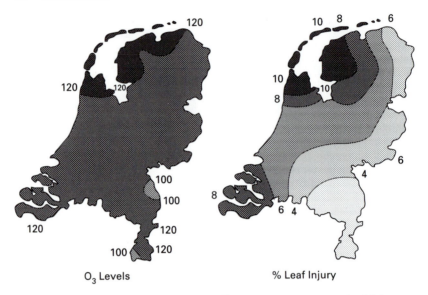

Fig. 1.6. Distribution of O_3 (in $\mu g\ m^{-3}$, left) and percentage leaf injury to tobacco (cv. Bel W_3) (right) across the Netherlands in 1979 (courtesy of Dr A. C. Postumus, RIPP Wageningen).

Non-plant bioindicators are also possible but little used. Leaf-eating insects, for example, distinguish and predate leaves from polluted plants rather than those from unexposed vegetation even though there are no visible differences between them. It is presumed that they are able to sense small differences in emitted odours from affected and unaffected leaves.

Similarly, the microbial population on the surface of leaves changes as a consequence of atmospheric pollution. In an interesting survey of atmospheric pollution around Dublin carried out by Irish schoolchildren, one of the tests followed the increase in numbers of 'pink yeasts' (actually fungi) after plating out the leaf imprints, which was more informative than the parallel lichen studies (see Fig. 1.5). Tests using microbes and insects appear simple but there is still little understanding of the fundamental mechanisms beneath them. Nevertheless, they demonstrate that useful and more detailed survey information can be gained without the widespread use of expensive monitoring equipment.

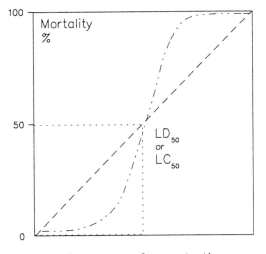

Fig. 1.7. The LD_{50} or LC_{50} (lethal dose/concentration that kills 50% of a group) usually assumes a sigmoidal curve which is then back-projected to a dose or concentration where no effect is thought to occur.

Thresholds and injury

One of the classical procedures of toxicology was to determine the median concentration which kills a certain proportion of organisms. The lethal concentrations or doses that kill 50% of the test animals in 48 h (i.e. the LC_{50} or LD_{50}) were frequently used during the testing of drugs, etc. Indeed, much legislation still requires that they be done before products enter commercial use. The concept of these tests was that these 50% values could then be applied to the midpoint of sigmoidal curves (Fig. 1.7) and then back-projected to give notional safe concentrations or doses that might be considered to be harmless.

Apart from providing a quick screening of likely hazards from an unknown toxicant, the logic of the rest of the interpretation is questionable. Death and sublethal harm are often unrelated. Even assuming all variation (which is usually considerable) has been removed from such experimentation, there is no direct means of extrapolating what happens, for example, in a rat to a human. Other variations on the same theme such as the MLD (minimum lethal dose) or MDNF (maximum dose never fatal) are little better.

As lethality is only a crude measure of toxicity, possible sublethal effects have also been monitored by toxicologists. The problem still remains – 'What change is harmful and what is within the normal range of homeostasis or internal self-regulation of cells?'.

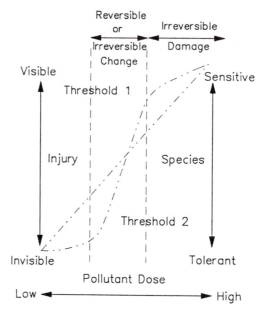

Fig. 1.8. Adaptation of the LD_{50} concept to interpret plant effects. The sigmoidal response suggests concentration or dose thresholds at which normal cellular regulation may turn into invisible injury or for this reversible change to become irreversible.

Many measures of sublethal toxicity (biochemical, mutagenic, carcinogenic, teratogenic) are used. For each potential toxicant there is a different type of response curve; some are linear and others are sigmoidal with clear thresholds. Each toxicant – especially air pollutants – differs in response. Most TLVs were derived by a combination of lethal and sublethal studies and have practical value for healthy individuals at work, but critical groups – the young, the old, the ill and the pregnant – are not covered because the original objectives were only intended to reflect the risks to healthy adults.

The original concepts developed by animal toxicologists were also used by plant pathologists to assess possible vegetational effects. Pathological change in plants is often revealed as visible leaf injury. However, adverse effects are not always visible. Unseen or invisible injury still occurs outside normal ranges of homeostasis and reduces growth. Usually, this is reversible but, if prolonged, it becomes irreversible and leads to visible injury (see Fig. 1.8).

Reversible and irreversible changes may occur together. Response differs even between individuals in the same population. Some tolerate change and adjust while others are sensitive and are unable to cope with the altered conditions.

The bioenergetic cost to a plant of normal homeostasis to cope with external influences such as atmospheric pollutants is reflected as a loss of energy which cannot then be converted into growth, reproduction, etc. The further along the response curve (Fig. 1.8) a plant or animal moves, the more energy has to be devoted to maintenance. At some point, a negative balance between income and expenditure is achieved which then leads to visible injury. In other words, there is no sharp distinction between normal homeostasis and so-called invisible injury. One counterbalances the other in a cellular cost/benefit assessment that is constantly updated as the surrounding environment changes.

Genetic adaptation

Industrial melanism

Evolution occurs rapidly when organisms are exposed to fresh stresses. In some cases this can be quite a quick process. For example, the ability of certain bacteria to circumvent antibiotics is a major cause for concern in the medical profession. By similar mechanisms, certain insect populations acquire immunity to insecticides and rats to rodenticides.

The classic case of atmospheric pollution darkening the barks of trees which then affects the colour of moths is an excellent example of genetic adaptation. This phenomenon of industrial melanism was first observed in 1848 around Manchester, UK when darker or *carbonaria* versions of the common peppered moth (*Biston betularia*) appeared. The overall process is relatively fast but starts slowly and enters an exponential phase, before slowing again. The frequency of dark (melanic) individuals in the population therefore shows a sigmoidal relationship with time. By 1895, over 95% of the individuals in the population around Manchester were black even though there is only one generation a year.

In the UK, there are now over 80 different species recorded as showing industrial melanism and many examples exist elsewhere in the world. The bulk of them have one feature in common. Most of them are insects which rest with their wings fully open on tree trunks or rocks and derive protection from predators (mainly birds) from colorations and patterns on their upper wings which closely resemble the background on which they rest (see Fig. 1.9). Many species also have blacker larval stages but this is controlled by a different set of genes.

Escape from predation by birds is a powerful agent of natural selection because a large number of those birds which hunt by sight actively search tree trunks for resting insects. Insects, on the other hand, do not necessarily alight on those trees which suit their colouring although they will move around once landed to accord better with their surroundings. There is thus a balance of wing (or larval) coloration of the population

Fig. 1.9. Examples of the common peppered moth (*Biston betularia*) of both the normal and *carbonaria* forms resting on a lichen-covered tree in rural Dorset, UK (left) and on a blackened tree in the industrial area surrounding Birmingham, UK (right). N.B. There are two specimens in each photograph! (Courtesy of the late Prof. E.B. Ford, FRS, Oxford.)

as a whole which is determined by the feeding activities of the birds in a particular environment. As the surroundings become blackened by deposition of particulates and enhanced sulphide formation on surfaces, the total gene pool (a collective expression for all genetic characters) of the population moves in favour of wing (or larval) blackening to correspond more closely with the surrounding environment in areas of atmospheric pollution. Whether this evolution takes place due to selective pressure by means of favouring small genetic variations already within a population or by the chance appearance of favourable mutants involving a sudden genetic change is still arguable. In every species showing industrial melanism, the trait is characterized by the spread of a dominant or semi-dominant mutation. If it were recessive then the offspring would not be melanic and any advantage would disappear. The rapid response to darker surroundings by virtue of adaptation therefore lies in the accumulation of genes that are at least partially dominant in their effect.

The original 1848 *carbonaria* specimen of the common peppered moth was a heterozygote (i.e. it carried genes for both melanic and normal colouring) and lighter in colour than both the heterozygotes and homozygotic 'modern' *carbonaria* forms that were trapped later and up to the present day. Crosses of 'modern' heterozygotes with 'normal' moths from non-industrial localities still produce the lighter-coloured 'original' *carbonaria* forms. In other words, the advantageous melanic mutant trait has been strengthened by the presence of other genes which have converted the partial dominance of the 'original' forms into the complete dominance of the 'modern' *carbonaria* type. It is, therefore, the whole gene pool within a population that forms the basis of successful adaptation.

Sensitivity and tolerance

Single gene mutations are rarely able to effect a complete response to an environmental stress. It is the accumulation of active gene combinations which gradually increases the effectiveness of a population to resist stress. However, certain individuals within a population may be particularly well endowed with these useful combinations and, as such, are worth further selection in breeding programmes.

In the case of tolerance to air pollution stresses, artificial selection of animals or plants is rarely made although the tolerances of tobacco, poplars and pines to O_3 have been improved. There must be considerable advantage in searching for (and exploiting) similar tolerances to air pollutants in other species. Throughout the world, atmospheric pollution is still increasing, especially in the developing countries. This means that certain ecosystems have only a short time to build up natural tolerance to stresses induced by air pollution.

Evolution of environmental tolerance is a well-known phenomenon in plants. By such means they survive and grow in the presence of high concentrations of heavy metals or salts, in conditions of prolonged drought, or in a wide variety of adverse conditions.

Resistance to pollution is defined as the ability to maintain growth and remain free from injury in a polluted environment. It need not be complete and takes into account two possibilities – stress avoidance, where pollutants are excluded, and stress tolerance, where detoxification mechanisms exist to counteract incoming pollution. In the case of resistance to the deposition of atmospheric pollutants, avoidance does not apply because plants have to open their stomatal apertures to function and, consequently, the principal resistance mechanisms are those of stress tolerance.

The best practical illustrations of tolerance to atmospheric stress have been shown by grasses, although it has been found in many other species. The hills around Manchester, UK have been experiencing atmospheric pollution for over 200 years – longer than anywhere else in the world. Consequently, the dairy farmers on the higher ground around Manchester, and probably those downwind from many other industrialized cities, have long known that any attempt at reseeding their pastures with commercial grass selections will be met with poor germination and stunted growth for some time afterwards. Only when sufficient previously-adapted strains of grass have had time to spread in from the surrounding pastures or older buried seed germinates are the previous conditions re-established. As an example of this Plate 1 shows the wide range of genetic tolerance within one cultivar of smooth meadow grass to SO_2 alone. Responses to NO_2 and mixtures of SO_2 plus NO_2 are similar, some individuals performing well while others die.

Such a pool of air pollution tolerance characteristics could be of great value to plants in other polluted localities, especially if they could be combined with desirable growth characteristics. Plant scientists can now accelerate such selections using both conventional plant breeding and genetic engineering. At Lancaster, for example, we have produced barley cultivars which tolerate high levels of atmospheric nitrogen oxides and even appear to use the nitrogen oxides as a source of nitrogen and to use less artificial fertilizer. Unfortunately, this is often at a cost because tolerance or detoxification mechanisms are introduced at the expense of overall yield. For agricultural crops, it all depends on how these costs are evaluated – whether they take into account all the environmental implications or they remain at the whim of politicians. For other species in the environment, natural selection is the arbiter.

Further reading

Ashworth, J.R. *Smoke and the Atmosphere.* Manchester University Press, 1933, Manchester.

Barry, R.G. and Chorley, R.J. *Atmosphere, Weather and Climate* (5th edn). Routledge, 1990, London.

Bishop, J.A. and Cook, L.M. (eds) *Genetic Consequences of Man Made Change.* Academic Press, 1981, London and New York.

Brimblecombe, P. *Air Composition and Chemistry.* Cambridge University Press, 1986, Cambridge.

Brunner, C.R. *Hazardous Air Emissions from Incineration.* Chapman & Hall, 1985, New York.

Dix, H.M. *Environmental Pollution.* John Wiley & Sons, 1981, Chichester.

Dunderdale, L. (ed.) *Energy and the Environment.* Royal Society of Chemistry, 1990, Cambridge.

Grefen, K., Reinisch, D.W. and Suess, M.J. (eds) *Ambient Air Pollutants from Industrial Sources: A Reference Handbook.* Elsevier Science, 1984, Amsterdam.

Harrison, R.M. (ed.) *Pollution: Causes, Effects and Control.* Special Publication No. 44, Royal Society of Chemistry, 1983, London.

Kettlewell, H.B.D. *The Evolution of Melanism.* Oxford University Press, 1973, London.

Mellanby, K. *The Biology of Pollution.* Studies in Biology No. 38, Edward Arnold, 1972, London.

Raiswell, R.W., Brimblecombe, P., Dent, D.L. and Liss, P.S. *Environmental Chemistry.* Edward Arnold, 1980, London.

Richardson, D.H.S. *Biological Monitors of Pollution.* Royal Irish Academy, 1987, Dublin.

Scientific American. *Chemistry in the Environment.* W.H. Freeman, 1973, San Francisco.

Steubing, L. and Jäger, H.J. *Monitoring of Air Pollutants by Plants: Methods and Problems.* Dr W. Junk, 1982, The Hague.

Treshow, M. (ed.) *Air Pollution and Plant Life.* John Wiley & Sons, 1984, Chichester.

World Health Organization. *Air Quality Guidelines for Europe.* WHO, 1987, Copenhagen.

Chapter 2

SULPHUR DIOXIDE

'That knuckle-end of England – that land of Calvin, oat-cakes and sulphur'[1]
by the Reverend Sydney Smith (1771–1845) while commenting upon Scotland.

Sources and cycling of sulphur

Homogenic changes

Both natural and human-induced (homogenic) emissions contribute to atmospheric levels of sulphur dioxide (SO_2) – a gas with an unpleasant, highly irritating odour when it occurs at concentrations greater than 1 μl l^{-1}. Global emissions of SO_2 from natural sources due to microbial activity, volcanoes, sulphur springs, volatilization from various surfaces including vegetation, sea-spray, and weathering processes (see next section) amount to 128 Mt of sulphur (S) per annum (a^{-1}). These are only 50% greater than those caused by human activities (currently 70 Mt a^{-1}) although more widely distributed over the surface of the planet and providing natural background levels of about 1 nl l^{-1} SO_2. However, biogenic emissions of reduced S-containing compounds from vegetation, wetlands, and oceans are larger and considered alongside H_2S in Chapter 4.

More than 90% of homogenic emissions of SO_2 arise over Europe, North America, India and the Far East. Emissions of SO_2 were highest (77 Mt S a^{-1}) during the late 1970s but have fallen over the last two decades as a result of emission controls, changes in the patterns of fuel consumption, and economic recession. Unfortunately, forecasts of increased coal consumption towards and beyond year 2000 (see Fig. 8.14) predict that global homogenic emissions of SO_2 will probably rise again by 30% upon present-day rates to 85 Mt S a^{-1}. Typical urban levels in developed countries currently range from 0.1 to 0.5 μl l^{-1} but concentrations in excess of these levels are still common despite legislation.

Burning of coal contributes by far the greatest proportion of the

---

[1] The last insult alludes to the pollution problems of Edinburgh which gave rise to the city's nickname, 'Auld Reekie'.

Table 2.1 Estimates of global exchange of sulphur in 1968 caused by the activity of man (mainly from Anderson, 1978)

Source	Consumption ($Tg\ a^{-1}$)	Recovery ($Tg\ a^{-1}$)
Voluntary		
Elemental sulphur extraction	18.0	18.0
Iron pyrites extraction	11.0	11.0
Involuntary		
Copper smelting	5.8	
Lead smelting	0.7	
Zinc smelting	0.6	
Oil refining	16.9	22.0
Oil burning	10.3	
Coal burning	46.0	
Natural gas burning	6.7	
Total	116.2	51.0

homogenic emissions of SO_2 (see Table 2.1). Elemental S and iron pyrites (FeS) are mined to produce fertilizers, etc., but these voluntary sources will be reduced in future in favour of reusing involuntary by-products such as gypsum ($CaSO_4$) from emission controls fitted to power stations. Such changes may take place because of economic factors rather than by legislative intervention.

The most obvious way to reduce SO_2 emissions is to use fuels with the lowest possible S content, and considerable advances have been made in the removal of S from fuels (including pulverized coal) before combustion. Consequently, the balance between the higher costs of such fuels and those associated with installation of additional pollution controls is now paramount.

Fluidized-bed combustion involves the burning of small particles of solid or liquid fuel at 500–700°C. These are held in a state of suspension by jets of air injected under pressure directed upwards. Steam and water pipes immersed in the 'bed' then transfer the heat to turbines very efficiently The potential amount of SO_2 emitted by combustion varies according to the S content of the fuel. However, limestone or dolomite particles can be added to a fluidized bed to reduce SO_2 emissions by up to 90% because the S is trapped in the form of sulphates of calcium or magnesium within the ash. This technique therefore adds to the amount of 'involuntary' S that may be recovered and allows coal of higher S content to be used in the presence of stringent emission controls. The lower temperature of combustion also reduces the emissions of nitrogen oxides (Chapter 3) and CO (Chapter 9) because of the air-saturating conditions. Of all the combustion technologies likely to be used for future power generation it still appears to have many advantages (Table 2.2).

Table 2.2   Efficiencies and emission problems (without flue-gas desulphurization) of power generation by different combustion processes

Type of power plant	Maximum thermal efficiency	Emissions		
		CO	SO₂	NO₂ + NO
Oil-fired	35–37%	trace	high	v. high
Natural gas-fired	35–37%	trace	low	v. high
Conventional coal-fired	35–40%	low	high	v. high
Gas turbine	18–28%	high	v. high	v. high
Fluidized bed	40–50%	trace	low	low

Most homogenic SO_2 emissions are made through tall stacks injecting the gases into the troposphere to heights at which the temperatures of the surrounding air and the rapidly cooling stack gases match. A discrete plume is then often carried away at that level. The behaviour of this plume is dependent upon the prevailing weather features and the contents may be dry-deposited close by or carried many hundreds of kilometres away from the source. However, when the plume encounters wet conditions, considerable amounts of the gaseous SO_2 (and nitrogen oxides) are removed by wet deposition.

Such acidic depositions, and their consequences, are separately considered in Chapter 5. However, four different dry removal mechanisms for SO_2 from the atmosphere may occur without the immediate presence of water droplets, but all of them generate acidic deposition because each forms SO_3 which then reacts with H_2O to form sulphuric acid (Reaction 2.1).

Monatomic oxygen may react with SO_2 (Reaction 2.2) within the plume or higher in the stratosphere to form SO_3, but a second mechanism based upon Reactions 2.3 and 2.4, which rely upon the extensive presence of free radical hydrocarbons, only rarely occurs. A third photochemical reaction may produce an energized form of SO_2 which then combines with O_2 to form SO_4. However, as this chemical compound has never been positively identified in the atmosphere, this light-driven acidification process is highly unlikely.

$$SO_3 + H_2O \Leftrightarrow H_2SO_4 \tag{2.1}$$

$$SO_2 + O + M^1 \Rightarrow SO_3 + M \tag{2.2}$$

$$O_3 + C_2H_4 \Rightarrow CH_2O_2^{\cdot} + HCHO \tag{2.3}$$

[1] M is a third entity, such as a surface or another gas molecule, which is capable of carrying away excess energy associated with the formation of the new bond.

$$CH_2O_2^{\bullet} + SO_2 \Rightarrow SO_3 + HCHO \qquad (2.4)$$

The fourth possibility, and by far the most important, involves highly reactive hydroxyl ($^{\bullet}$OH) free radicals (see Appendix 1) formed from O_3 (see Chapter 6) or monatomic oxygen (Reactions 2.5 and 2.6). These radicals react readily with SO_2 (Reaction 2.7) to form $HSO_3^{\bullet}$ radicals which are then oxidized to peroxyl radicals ($HO_2^{\bullet}$) and SO_3 (Reaction 2.8). An equivalent sequence of reactions may also form nitric acid from NO_2 (see Chapter 3). The predominance of this fourth mechanism of acidification involving $^{\bullet}$OH radicals means that warm, bright summer days, which favour O_3 and monatomic oxygen formation (and hence $^{\bullet}$OH radicals), promote dry-phase acidification, while wet-phase reactions (see Chapter 5) are more important in winter.

$$O_3 + light \Rightarrow O + O_2 \qquad (2.5)$$

$$O + H_2O \Rightarrow 2^{\bullet}OH \qquad (2.6)$$

$$^{\bullet}OH + SO_2 + M^1 \Rightarrow HSO_3^{\bullet} + M \qquad (2.7)$$

$$HSO_3^{\bullet} + O_2 \Rightarrow HO_2^{\bullet} + SO_3 \qquad (2.8)$$

Figure 2.1 shows a number of the interrelationships between wet and dry mechanisms of SO_2 removal, but the corresponding wet-phase reactions are discussed in detail in Chapter 5.

Natural changes

Sulphur dioxide is emitted into the atmosphere by volcanoes and hot springs along with other gaseous forms of S such as H_2S (see Chapter 4). In terms of overall cycling of S, however, weathering of S-containing minerals like gypsum ($CaSO_4$) is more important. Such weathering is accelerated by acidification processes caused by microbial activity or by atmospheric pollution, and most of this abraded S slowly enters soils to act as a plant nutrient. In agricultural areas, where soils are naturally deficient in S, sulphate would normally be added artificially (along with nitrogen, phosphate and potassium) in fertilizers. However, in most regions of Europe and the eastern USA subject to SO_2 pollution, such treatments are quite unnecessary because amounts of S falling on the land by wet and dry atmospheric depositions are nearly five times that achieved by weathering, even in areas well endowed with rocks of a high S content.

Sea-spray is an alternative route by which S compounds may enter the atmosphere. Such particles may remain suspended for long periods because their droplet sizes are often very small. Most of this S, in the

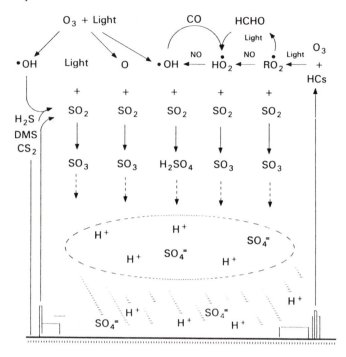

Fig. 2.1. Gas-phase oxidation and removal mechanisms for atmospheric SO_2 (courtesy of Drs Cox and Penkett, AERE, Harwell, UK, and D. Reidel Publ. Co., Dordrecht, The Netherlands).

form of sulphate, stays above the oceans and returns to the sea but as much as 10% may be carried inland.

The global significance and the interrelationships that exist between homogenic, natural, and microbial interconversions of S between the buried S reserves of the world, the surface of the land, the atmosphere, and the oceans are illustrated in Fig. 2.2. The important feature to note is that the amount of SO_2 generated homogenically is five times that generated from weathering. Furthermore, more of this atmospheric pollutant returns to the oceans than to the land because of the greater surface area of the sea. What is not so evident, but is equally important, is that the temporary reduction (in geological terms) of S reserves by homogenic activities eventually ends up as deposits on the floor of the oceans as new S reserves.

Microbial changes

Land plants, seaweeds and microbes volatilize S into the atmosphere in various forms. Vegetation, for example, releases H_2S into the atmosphere

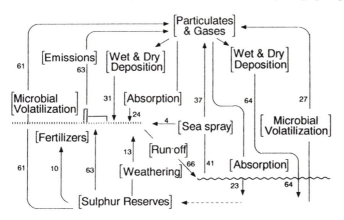

Fig. 2.2. Fluxes of sulphur in different forms between various global compartments (expressed in gigamoles of sulphur per annum).

as a mechanism to eliminate excess S (see Chapter 4) but there are several microbial exchanges of S between the biosphere and the environment (Fig. 2.2). This stems from the wide variety of different chemical forms that S may take up. By understanding these differences, a balanced appreciation of the effects of SO_2 on living systems may be gained.

Compounds containing S, like those with nitrogen, are unusual in that they take up a large number of oxidized or reduced states. Some of these different compounds are illustrated horizontally in Fig. 2.3. These compounds also differ energetically from each other – a concept best expressed in terms of their electrochemical potentials (shown vertically in Fig. 2.3). As electrons flow downwards (in Fig. 2.3) from electron donors (which become oxidized) to electron acceptors (thereby reducing them) they release energy. Conversely, in order to make electrons flow in the opposite direction (upwards) energy has to be put in. The analogy with a series of waterfalls in an ornamental garden driven by an electric pump is most appropriate.

Biological systems, especially microbial ones, exploit these energy differences. Chemoautotrophic bacteria such as *Thiobacillus*, *Thiovulum* and *Thiospirillopsis*, for example, derive energy by oxidizing S and sulphide to sulphite and sulphate to form adenosine triphosphate (ATP), the dominant energetic currency of living systems, which can then be used to assist with the biosynthesis of virtually every other compound that living organisms require (Fig. 2.4, Route A). These chemoautotrophic bacteria tend to occur on the surface of stagnant ponds at the boundary of anaerobic (oxygen-lacking) conditions (the source of the sulphide) and aerobic (oxygen-rich) conditions where O_2 is converted to H_2O.

Some chemoautotrophs are responsible for immense damage to build-

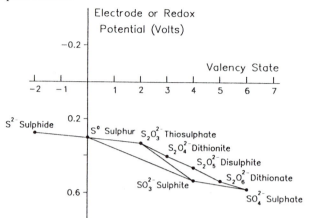

Fig. 2.3. Plot of redox potentials against valency states for various sulphur compounds. The vertical axis has been inverted because electronegative compounds contain more energy and naturally donate electrons to electropositive compounds lower down. A scale of electrode or redox potentials represented in this way gives a better appreciation of the relative energies within different compounds.

ings because they tolerate high acidity (below pH 1) caused by the sulphuric acid they release. Sulphuric acid released by *Thiobacillus concretivorus*, for example, steadily attacks the $CaCO_3$ of concrete, mortar or limestone and rots away such structures. Damage to sewers is particularly severe because the anaerobic conditions, which prevail for long periods, cause the build-up of reduced forms of S. These are then followed by surges of oxygenated water from downpours of rain which are ideal conditions for the growth of *Thiobacillus* and its concrete-rotting exudations. More useful chemoautotrophic soil bacteria, however, oxidize adsorbed SO_2 to sulphate which can then be taken up by plant roots. Similarly, the H_2S released by anaerobic soils is detoxified by chemoautotrophic bacteria nearer to the soil surface before it can escape into the atmosphere to cause problems (see Chapter 4).

A number of microbial processes also cause the reduction of oxidized forms of S by a process known as 'assimilatory photosynthetic sulphur reduction' (Fig. 2.4, Route E). Alternatively, a trapped food source such as glucose may be used to generate the necessary energy and electrons by an anaerobic process known as 'assimilatory heterotrophic sulphur reduction' (Fig. 2.4, Route D). Most fungi, yeasts and aerobic bacteria synthesize S-containing amino acids in this way. Consequently, these two types of assimilatory S reduction are the major routes by which sulphate from the atmosphere is adsorbed and trapped by the biosphere.

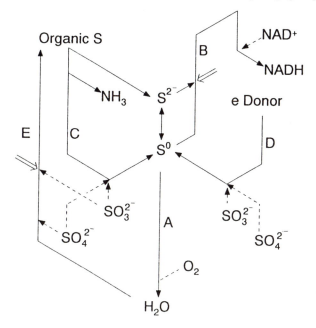

Fig. 2.4. Electron flow (solid lines) to and from native sulphur (S^0) and sulphide (e.g. H_2S) during **A**, chemoautotrophy; **B**, photoautotrophy; **C**, desulphurylation (associated with putrefaction); **D**, respiratory or dissimilatory sulphur reduction or sulphur respiration; and **E**, assimilatory sulphur reduction. The open arrows indicate light-driven events during **B** and **E**. Route **D** is also used for assimilatory heterotrophic sulphur reduction not primarily to derive energy (ATP, etc.) but to form reduced sulphur for the synthesis of S-containing amino acids.

Inevitably, more S is trapped than is needed and volatilization of the excess occurs mainly as H_2S (see Chapter 4). The uptake of S by most assimilatory S-reducing microbes usually balances that required for nutritional purposes and only a small proportion of sulphide is released or accumulated. However, the same cannot be claimed for another anaerobic process known as 'dissimilatory or respiratory sulphur reduction' (Fig. 2.4, Route D) which is carried out by bacteria in the genera *Desulfovibrio* and *Desulfotomaculum*. These organisms, found in waterlogged soils or mud, stagnant ponds, sewage pipes, and the rumens of grazing animals, take in large amounts of sulphate or sulphite and reduce them to elemental S or sulphides. These products are then released around these organisms which use only a small fraction for the synthesis of S-containing amino acids. Usually, dissimilatory S reduction uses a reduced carbon source as a source of electrons and the energy is trapped as ATP. As sulphate is used as a terminal electron acceptor in the absence of oxygen (Fig. 2.4, route D), this process is also often called 'sulphur respiration'.

By comparison to assimilatory sulphur reduction of plants, fungi and aerobic bacteria, S respiration plays only a small part in removing S from the atmosphere because these organisms live in an anaerobic environment. However, they do cause environmental problems, from the smells they cause when their habitat is disturbed and from their involvement with the corrosion of underground metal pipework. Normally, when iron pipes are in contact with water, some of the metal dissolves and H_2 is formed (Reaction 2.9). If this is not removed, an opposing electrochemical potential is formed (i.e. it becomes polarized) and the iron stops dissolving. Some dissimilatory bacteria, however, possess the ability to utilize H_2 as a source of electrons to reduce the sulphate (Reaction 2.10). This depolarizes the system and permits corrosion to continue (Reaction 2.11).

$$Fe + 2H_2O \Rightarrow Fe^{2+} + 2OH^- + H_2 \qquad (2.9)$$

$$SO_4^{2-} + 4H_2 \Rightarrow S^{2-} + 4H_2O \qquad (2.10)$$

$$S^{2-} + 2Fe^{2+} + 4H_2O \Rightarrow FeS + Fe(OH)_2 + 2OH^- \qquad (2.11)$$

Reactions of sulphur dioxide in water and tissue fluids

Sulphur dioxide molecules contain no unpaired electrons, and therefore are not free radicals, but readily dissolve in water to form sulphite (SO_3^{2-}) and bisulphite (HSO_3^-) ions (Reaction 2.12). The remaining unionized SO_2 may be considered as either sulphurous acid (H_2SO_3) or a dissolved form of SO_2 although little of this unionized SO_2 exists in solution. The relative proportions of bisulphite and sulphite in solution are more important because the dissociation equilibrium between the two has a pK_a (see Appendix 2) of 7.18 which means that around neutrality (i.e. similar to that in biological tissues) roughly equal proportions exist (see Fig. 2.5).

Both sulphite and bisulphite have a lone pair of electrons on the S which means that attack on electron-deficient sites in other molecules is strongly favoured. Consequently, they are readily oxidized by a series of overlapping reactions which involve the formation or consumption of free radicals (Reactions 2.13–2.18). The additional presence of a metal such as manganese is vital for some of these rapid oxidations (Reactions 2.13 and 2.14) and for the production of superoxide ($^{\cdot}O_2^-$) by two different reactions (Reactions 2.13 and 2.15).

Other free radicals of S have also been detected in addition to the bisulphite radical ($HSO_3^{\cdot}$) implied by Reactions 2.13–2.17. For example,

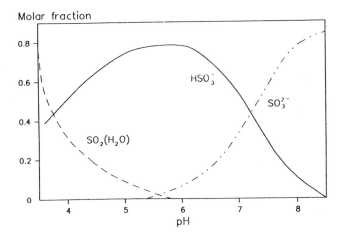

Fig. 2.5. Relationship between amounts of sulphite, bisulphite, and dissolved SO_2 at different pH values.

the actions of light or reducing agents are capable of forming sulphoxyl radicals ($SO_2^{\cdot-}$) which persist for much longer than most free radicals. These are capable of donating electrons to electron-deficient sites and oxidizing sulphite to sulphate (Reactions 2.19 and 2.20).

$$SO_2 + H_2O \Leftrightarrow H_2SO_3^- \Leftrightarrow H^+ + HSO_3^- \Leftrightarrow H^+ + SO_3^{2-} \quad (2.12)$$

$$HSO_3^- + O_2 \overset{Mn^{2+}}{\Rightarrow} HSO_3^{\cdot} + {}^{\cdot}O_2^- \qquad (2.13)$$

$$SO_3^{2-} + O_2 + 3H^+ \overset{Mn^{2+}}{\Rightarrow} HSO_3^{\cdot} + 2{}^{\cdot}OH \qquad (2.14)$$

$$HSO_3^{\cdot} + O_2 \Rightarrow SO_3 + {}^{\cdot}O_2^- + H^+ \qquad (2.15)$$

$$HSO_3^{\cdot} + {}^{\cdot}OH \Rightarrow SO_3 + H_2O \qquad (2.16)$$

$$2HSO_3^{\cdot} \Rightarrow SO_3 + SO_3^{2-} + 2H^+ \qquad (2.17)$$

$$SO_3 + H_2O \Rightarrow SO_4^{2-} + 2H^+ \qquad (2.18)$$

$$SO_2 + OH^- \overset{\text{light or reducing agents}}{\Rightarrow} SO_2^{\cdot} + {}^{\cdot}OH \qquad (2.19)$$

$$SO_2^{\cdot} + O_2 + H^+ + e \Rightarrow HSO_4^- \qquad (2.20)$$

$$SO_3^{2-} + [O] \overset{\text{sulphite oxidase}}{\Rightarrow} SO_4^{2-} \qquad (2.21)$$

It is undesirable that uncontrolled levels of free radicals should rise in living tissues because so many other troublesome reactions may take place. As both Chapters 3 and 6 emphasize, most tissues have

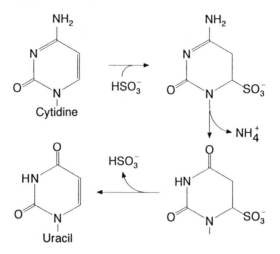

Fig. 2.6. Conversion of cytidine to uracil in DNA by bisulphite which is responsible for some of the mutagenic properties of this anion.

comprehensive free radical scavenging systems which mop up free radicals as soon as they are formed. The case of sulphite is no exception. Rather than rely upon the above reactions to oxidize sulphite to sulphate, nearly all organisms have a molybdenum-containing enzyme called sulphite oxidase which also ensures the rapid removal of sulphite (Reaction 2.21).

Reactions of sulphite with biomolecules

There are a large number of possible interactions of sulphite with other critical biochemical compounds. Reactions 2.12–2.20 demonstrate that a wide variety of free radicals (HSO_3^-, $^{\bullet}O_2^-$, $^{\bullet}OH$, and SO_2^-) may be formed as sulphite is oxidized and this has a number of consequences in biological tissues. The most likely are chain cleavage of DNA, oxidation of the double bonds in fatty acids within membranes, and the conversion of the amino acid methionine into methionine sulphoxide (see Chapter 6).

However, bisulphite and sulphite are both highly reactive nucleophiles in their own right. Indeed, the efficiency of some of their reactions with organic compounds forms the basis of food preservation using SO_2. Amateur wine-makers are very familiar with the sterilizing capabilities of sodium sulphite tablets. In high concentrations, sulphite is also mutagenic by virtue of the conversion of cytidine to uracil (Fig. 2.6) within DNA which causes guanine–cytidine base pairs to turn into adenine–thymidine

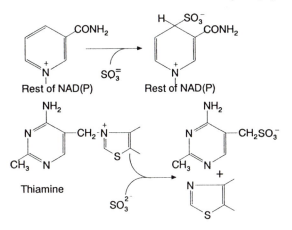

Fig. 2.7. Attack of sulphite on the nicotinamide ring of pyridine-linked dehydrogenases or during the destruction of the vitamin, thiamine.

associations because uracil reads the same as thymidine upon replication of nucleic acids.

The sterilizing effect of SO_2, however, also resides in other disruptive reactions within microbes, killing them immediately. Chief among these is the attack on disulphide bridges within enzymes and structural proteins (Reaction 2.22). This inactivates critical reactive centres of hydrolytic enzymes in microbes, or the food itself, thus retarding structural breakdown and undesirable microbial contamination.

$$R-S-S-R' + HSO_3^- \Rightarrow R-S-SO_3^- + R'-SH \qquad (2.22)$$

Other important nucleophilic attacks of SO_2 also take place on important vitamin-linked compounds such as NAD^+, $NADP^+$, FAD, FMN, pteridines, folate, tryptophan and thiamine. The last of these is a particular problem for the food industry because the attack of sulphite on thiamine (Fig. 2.7), a vitamin of the B complex, is so efficient that some SO_2-preserved baby foods have to be fortified with thiamine by manufacturers to offset the possibility of thiamine deficiency.

During exposure to atmospheric SO_2, however, levels of sulphite are much too low to allow a similar reaction to those that occur during food preservation when one of the reactants (SO_2) is in vast excess. One reaction with SO_2 that may produce low levels of highly toxic products is the formation of α-hydroxy-sulphonates (Reaction 2.23). Such compounds are known to interfere, for example, with photorespiration (see Chapters 8 and 9). Furthermore, attack of sulphite on ring systems of NAD^+ (Fig. 2.7), $NADP^+$, FMN and FAD to form similar derivatives may occur. As these cofactors are vital for enzyme function, the inhibitory consequences of α-hydroxy-sulphonates disabling enzymes in this way could be an important cellular explanation of SO_2-based injury.

$$R-CHO + HSO_3^- \Rightarrow R-CH(OH)-SO_3^- \qquad (2.23)$$

Plant effects

Stomatal access

The first barrier encountered by a gaseous pollutant before reaching a leaf is a layer of still air above the leaf surface which imposes a 'boundary layer resistance' to the entry of atmospheric pollutants. Unstirred air layers on both sides of a leaf therefore represent an important resistance (see Fig. 2.8) to the access of SO_2. Movement across such a layer is achieved by diffusion of gas molecules in response to differences in concentration. The boundary layer resistance also varies with wind speed and leaf properties such as size, shape and orientation. As wind speed increases, resistance to the inward movement of pollutant molecules falls and uptake increases; in this way, increased wind speed can facilitate the uptake of pollutant. Moreover, the boundary layer is generally thinner at the edges of a leaf than at the centre, which accounts for increased pollutant damage often seen at margins of leaves, especially on grasses and cereals.

Although epidermal cells occupy a greater proportion of a leaf surface than do stomatal pores, the waxy cuticle covering them offers a greater penetration barrier to most pollutant gases. Indeed, this cuticular resistance (Fig. 2.8) is much greater than through the stomatal pores and into the internal air spaces beneath. Consequently, cuticular resistance is often ignored in any consideration of pollutant gases entering a leaf. However, SO_2 deposited on wet leaf and stem surfaces may also dissociate and react with cuticular waxes. In these circumstances, a certain amount of SO_2 may then enter leaves by penetrating the damaged cuticle even though most atmospheric SO_2 still enters through stomata. When the cuticle remains damp for long periods, especially at night or in the winter, the deposition rate of SO_2 onto a leaf surface increases despite the fact that the stomata are closed. This process has additional consequences for evergreens which retain their leaves and needles over winter (see Chapters 5 and 10).

For stomata to open, guard cells of the stomata have to be turgid which causes the gap between them to widen. In most plants, this occurs during the day as normal exchanges of water and CO_2 take place. As a result, uptake of atmospheric gases like SO_2 into plants takes place

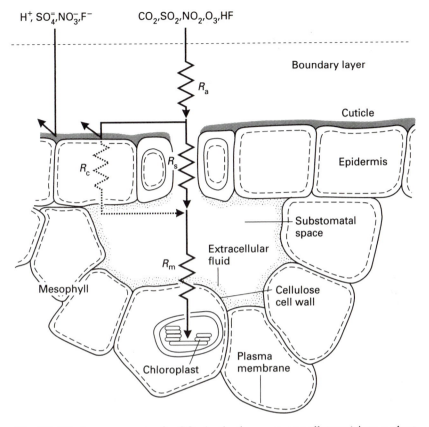

Fig. 2.8. Likely access routes for SO_2 (and other gaseous pollutants) into a plant leaf. The layer of still air or boundary layer imposes a resistance to flow (R_a) which depends on a number of factors including wind speed. Access then may be either restricted or enhanced by the degree of opening of the stomata (R_s) or by penetration of the cuticle and epidermal layers (R_c). In the cases of SO_2 and O_3, entry through the stomata is thought to be the major route of access, but in the case of nitrogen oxides a small fraction may cross the cuticle. The mesophyll resistance (R_m) consists of a number of different components including the still air of the substomatal space, the extracellular fluid, the cellulose cell wall, the plasma membrane, the cytoplasm, and any organelle envelope membranes before major sites of reaction are reached.

during the day. In very dull conditions, stomatal opening is reduced and, therefore, rates of SO_2 uptake are also reduced. At night, however, when stomata are closed, rates of removal of SO_2 (or O_3) by plants approach those of other inert surfaces. Consequently, those changes in stomatal aperture, which provide a plant with mechanisms to control the movement of CO_2 into leaves and H_2O outwards, also influence pollutant gas uptake. Stomatal behaviour, frequency and distribution are therefore important factors which affect the amount of pollutant entering a plant.

At high humidities, guard cells respond differently to the presence of SO_2 as compared to other pollutants. SO_2 causes guard cells to become more turgid, as well as sometimes damaging the surrounding subsidiary cells, which opens stomata wider and allows more polluted air to enter. If the pollutant is removed quickly, these effects are reversible (although less so if prolonged) but it means that, in warm and wet conditions, access of SO_2 is made easier. In dry conditions, when humidities are low, the converse is true. Stomatal pores open less and hence the resistance to pollutant entry is greater. However, this closure may also be due to other drought-induced protective mechanisms rather than to a direct effect of SO_2.

The opening response in high humidities due to SO_2 has long-term implications, especially in conifers, if increases in water loss occur. The mechanism by which SO_2 causes these turgor changes in guard cells has not been fully investigated but it is presumed that pH changes and enhanced levels of bisulphite and sulphite from SO_2 disturb the fluxes of potassium, calcium, chloride, malate and protons between guard cells and their subsidiary cells which are also involved in the stomatal response. Calcium-regulated channels through the guard cell membranes are part of the stomatal closing responses and sulphite is believed to interfere with this calcium-dependent mechanism.

Furthermore, SO_2 is nearly 30 times as soluble as CO_2 in aqueous fluid. This means that SO_2 uptake takes place mainly on the lower inner surfaces of the stomatal guard cells, where most (77–90%) of the transpired H_2O is lost, rather than from the mesophyll extracellular fluid which contributes only 10–23% of H_2O loss but 85% of CO_2 uptake (see also Fig. 2.8). This is largely because removal mechanisms of CO_2 are associated with photosynthetic carbon fixation. Uptake of SO_2 close to the stomata also means that the immediate effects of SO_2 are focused in this area. Indeed, SO_2 injuries both to the plasma membranes of guard cells and to the transport cells loading the vascular elements of leaves (companion cells in angiosperms, Strasburger cells in gymnosperms) are especially important. This means that increased levels of acidity and other ions (e.g. sulphite and sulphate) occur at these damaged locations and together they can have a detrimental effect upon the ability of a plant to move and use H_2O and photosynthate.

Internal resistance and buffering

Just as access of SO_2 across the boundary layer and through the stomatal pores can be treated as variable resistance (Fig. 2.8), the movement of SO_2 (and its products sulphite or bisulphite) to sites of action across the substomatal cavity in the extracellular fluid on the internal mesophyll cell walls can be interpreted as offering a resistance.

In fact, this 'mesophyll resistance' (R_m) is neither constant nor negligible and varies between species and even between cultivars. Indeed, pollution tolerance or sensitivity to SO_2 within a species is partly due to differences in mesophyll resistance. Moreover, the mesophyll resistance is not a simple diffusive resistance because the SO_2 and its derivatives have to cross the cellulose cell wall, the cell (or plasma) membrane, some of the inner cytoplasm, and even double envelope membranes of organelles before reaching likely targets such as chloroplasts. As they pass, they disrupt membranes, change the ionic environment of the cytoplasm, and cause a number of other secondary changes.

The behaviour of SO_2 in water and tissues has already been discussed, but the critical factors are the buffering capacities in the extracellular fluid, the cytoplasm, inside the chloroplasts, etc. Cell buffering (i.e. the ability to counteract a change of pH from an optimal value) is carried out by means of a wide variety of molecules, both large (often proteins) and small (usually amino acids), which have free carboxyl ($-COO^-$), phosphate, amino ($-NH_3^+$) or sulphydryl ($-SH$) groups. These are momentarily overloaded if high doses of SO_2 are suddenly experienced but a similar dose at low concentration for a longer period presents no problem. It has been claimed that the tissues of plants exposed to SO_2 have a lower buffer capacity than tissues of unexposed plants. If, however, pH should change in certain parts of the cell (e.g. the chloroplast) the consequences for cellular metabolism could be considerable.

As discussed earlier, in the section on reactions of SO_2 in water and tissue fluids, a variety of different dissociation equilibria exist for the products of SO_2 in solution. Normally, unionized weak acids enter cells and organelles easily because they cross membranes freely. On the other hand, fully dissociated strong acids produce anions which are unable to cross membranes. Instead, embedded proteins in membranes called pumps or translocators actively pass these charged species from one side of a membrane to another. The phosphate translocator spanning the inner chloroplast envelope is one such example. This transport protein behaves as an antiporter because it exchanges inorganic or organic phosphate going one way for a proton or a different organic form of phosphate coming the other way. Moreover, it also allows sulphite or sulphate to travel across the membrane in a similar 'piggy-back' fashion. Consequently, strong acids (e.g. sulphurous and sulphuric) gain access to cellular compartments more readily than passive diffusion would normally allow.

Algal studies have shown that there is a positive linear relationship between cytoplasmic pH and external pH, but in higher plants the ability to resist and control intracellular gradients of acidity and alkalinity is better. In lichens, however, control of pH by a 'pH-stat' mechanism is very primitive and relies extensively, but not entirely, on the inherent buffering capacity. It is commonly believed that lichens as a whole are especially sensitive to SO_2 because of their scarcity in heavily polluted areas. This is true for only some lichens. More importantly, polluted regions of northern Europe and the USA often have high humidities which encourage luxuriant growth of lichens so the likelihood of damage is much greater. There is also a greater chance of other biotic and abiotic stresses within these areas, which limit the growth of certain lichens. Therefore, taken as a group, lichens are no more sensitive to SO_2 pollution than higher plants, although they have proved most useful as biological indicators of the long-term presence of adverse atmospheric pollution conditions within a particular locality or region (see Chapter 1).

In higher plants, the mechanism of the pH-stat is complicated and only partly understood. Buffering capacity plays a part but proton pumps exist in membranes to allow protons to be transferred from one cellular compartment either to another or to the outside. The proton pump of the cell (or plasma) membrane is unaffected by either sulphate or nitrate (from NO_2 – see Chapter 3) but the proton pump of the tonoplast (the membrane bounding the vacuole of plant cells which is often filled with acidic storage or waste materials) is strongly inhibited by nitrate and, to a lesser extent, by sulphate. In other words, if anions like nitrate or sulphate gain access to the cytoplasm from extracellular fluid then the normal pH-stat mechanism of regulation which increases export of protons to the acidic vacuole, is hampered. Consequently, the only effective means of pH control is for the plasma membrane translocators to pump harder outwards against a concentration gradient already made worse by atmospheric pollution coming in. This means more energy is needed to resist pH changes and less is available for growth.

Sulphur metabolism

As mentioned earlier, higher plants and algae use photosynthetic electron flow to reduce sulphate to sulphydryl groups ($-SH$) as in the synthesis of S-containing amino acids by a process known as 'assimilatory photosynthetic sulphur reduction'. Reduction of sulphate occurs by two metabolic pathways (Fig. 2.9) – one which involves the 'active sulphur cycle' and one which is similar to nitrate reduction (see Chapter 3).

Sulphur reduction involves the formation of sulphite from sulphate which is then reduced to sulphide. However, the situation differs across the cell because sulphite is rapidly oxidized to sulphate in the chloroplast by light and by sulphite oxidase in the mitochondria. Superoxide ($^{\cdot}O_2^{-}$) is involved in photo-oxidation but low levels of sulphite can donate electrons to Photosystem II. The $HSO_3^{\cdot}$ free radical which is then formed accepts electrons from Photosystem I to become sulphite again. This means that low levels of sulphite exist naturally in close proximity

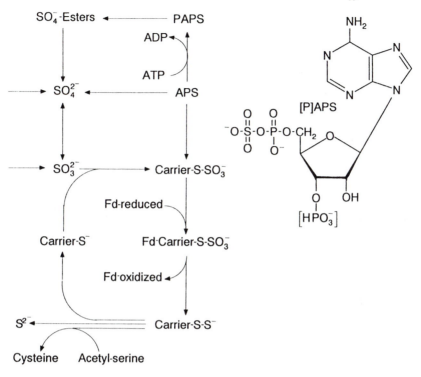

Fig. 2.9. Natural mechanisms for the oxidation and reduction of sulphate and sulphite in biological tissues. To the left is the so-called 'active sulphur cycle' involving the sulphonated nucleotides, adenosine-5'-phosphosulphate (APS) and adenosine-3-phospho-5'-phosphosulphate (PAPS), the structure of which is on the right. The sulphate on the former (i.e. with H− instead of an extra phosphate in brackets) is also reduced to sulphide by a series of reactions (lower half) involving the iron–sulphur protein, ferredoxin (Fd).

to the chloroplast membranes and are repeatedly recycled by this futile oxidative process while, at the same time, reduction to sulphydryl groups also takes place. Only when the input of sulphite from SO_2 fumigation becomes excessive does this delicate balance fail and problems arise from enhanced levels of free radicals which are not held in check by scavenging processes.

Reduction of sulphite in chloroplasts produces sulphide and H_2S (see Chapter 4) as well as sulphydryl groups (−SH). Moreover, release of H_2S by leaves in the light but not in the dark is one of the means by which plants rid themselves of excess S brought about by SO_2 exposure.

Some unreduced sulphite is oxidized to sulphate and taken up by the 'active sulphur cycle' (upper part of Fig. 2.9). This involves the formation of two compounds related to ATP, adenosine-5'-phosphosulphate (APS) and adenosine-3-phospho-5'-phosphosulphate (PAPS) which are involved in the sulphation of many organic molecules (mainly lipids). The remaining sulphite is reduced and converted into S-containing amino acids such as cysteine (lower part of Fig. 2.9). Once formed, this amino acid is the basis for a series of interconversions which allow the formation of all the other S-containing amino acids, including methionine, required by plants, fungi and most microbes.

The situation is reversed in humans. Most monogastric animals require methionine in their diets because they lack most of the biosynthetic enzymes between inorganic S and either cysteine or methionine. Therefore, in non-ruminants, emphasis is placed upon breakdown of methionine first to cysteine, etc., and ultimately to NH_3 and sulphite. The latter then requires immediate oxidation to sulphate by sulphite oxidase before elimination through the kidneys.

Ruminant animals differ from humans and other monogastric animals because they can use inorganic S from herbage as their only source of S. This is due to the fact that the microbes of the rumen (which precedes the true stomach – the abomasum) either reduce sulphate to sulphide or remove S from S-containing amino acids by desulphurylation (see earlier) and then resynthesize their own S-containing amino acids from sulphide. The bacterial proteins and amino acids released by digestion in the abomasum at pH 1 are then absorbed into the ruminant bloodstream through the intestinal wall. This means that grazing animals feeding on herbage which is deficient in S-containing amino acids require only inorganic sulphate to be added to their diets. In practice, this is rarely needed because, even on soils which would be naturally low in S, atmospheric deposition of SO_2 is usually more than adequate to compensate for this deficiency.

Chloroplastidic damage

Inhibition of photosynthesis is frequently thought to be one of the first effects of SO_2 upon plants. Consequently, the chloroplast is often regarded as the primary site of many disturbances caused by SO_2 or its products in aqueous solution. The pH of the chloroplast stroma is generally much greater than pH 7 (nearer pH 9 in the light) and this favours the formation of sulphite ions at the expense of bisulphite when S ionizes in solution (Fig. 2.5). Consequently, the effects of sulphite are often considered to be a reflection of the mode of action of SO_2 within chloroplasts. Indeed, ultrastructural studies have shown a swelling of the lumen spaces within the thylakoids to be one of the first effects of SO_2. Initially, this swelling is reversible although the time taken for recovery is proportional to dosage. Such swelling is indicative of ionic disturbances and cellular acidification mentioned earlier.

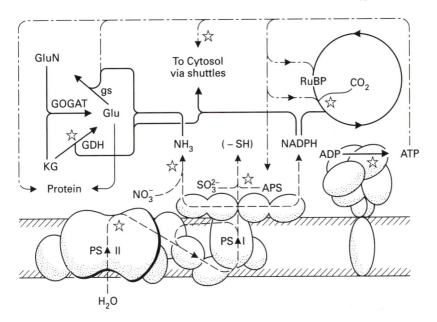

Fig. 2.10. Outline of various events within chloroplast membranes and stroma associated with photosynthesis. The flows of electrons through the photosystems are used for a variety of different purposes (i.e. reduction of $NADP^+$, nitrite, sulphite or APS) but they also induce a pH gradient across the membrane (not shown) which is harnessed by the coupling factor particles to form ATP (lower right). The uses of ATP are many but a number of the most important are shown in this diagram, all of which may be detrimentally affected by SO_2 and its products (in Unsworth and Ormrod (eds), 1982).

Figure 2.10 summarizes the important events of photosynthesis in chloroplasts. It includes the biochemistry of CO_2 fixation and implicates S and nitrogen metabolism. All these events are dependent upon electron flow from water induced by light falling on the photosystems (PS I and PS II), which creates a proton gradient across the internal thylakoid membranes causing the stroma (where the CO_2 is fixed) to become more alkaline. This alkalinization enhances the activity of enzymes fixing CO_2 and, consequently, any increase in acidity due to SO_2 pollution is likely to be significant. For example, direct effects of SO_2 have been demonstrated upon the enzyme of CO_2 fixation, ribulose-1,5-bis-phosphate carboxylase/oxygenase (given the 'breakfast cereal manufacturer' acronym of RubisCO), which requires a pH change in the stroma from pH 7.5 to pH 9 to become active. It has been calculated that a reduction of chloroplast stroma by 0.5 of a pH unit will cause a 50% decrease in net photosynthesis, so even a small change could be significant to growth of polluted plants over a long period.

Figure 2.10 also emphasizes some of the biosynthetic demands placed on ATP availability both inside and outside chloroplasts. Detrimental effects of SO_2 and its products upon all of the processes leading to photophosphorylation (i.e. the formation of ATP from ADP and orthophosphate by light) have been detected. Consequently, impaired operation of the photosystems, interference with electron flow, leakiness of the membrane to protons, failure to utilize the proton gradient, or inhibition of the sites of phosphorylation by SO_2 pollution, all go on to cause a reduction of the rates of protein and carbohydrate synthesis because these processes require ATP. There is good support for this 'ATP deprivation' concept because reductions in ATP pool sizes in pines, for example, correlate with increases in atmospheric SO_2 levels. Similar reductions in the ability to form ATP have also been detected in plant mitochondria.

The integrity of the photosynthetic membranes to maintain a proton gradient across the thylakoid membrane (which is generated by electron flow and harnessed by the coupling factors to make ATP – see right-hand side of Fig. 2.10) is also disturbed by the presence of sulphite, etc., especially in the presence of other anions like nitrite. Sulphite and nitrite together promote additional free radical formation which causes membranes to become 'leaky' with respect to protons. Consequently, insufficient ATP is produced which then leads to reduced growth. This interaction is discussed further in Chapter 11.

Long-term injury

One of the commonest types of visible injury caused by SO_2 is chlorosis (whitened areas of dying tissues where the pigments are breaking down). Plate 3 shows typical signs of chlorosis on lucerne (*Medicago*) due to SO_2. Much early research focused on describing obvious signs of visible injury (i.e. chlorosis, necrosis, early leaf fall, and increases in numbers of dead leaves, etc.) but there are hundreds of publications in which exposure to SO_2 is claimed to exert an effect (usually adverse) upon the growth of crops, trees, lichens and mosses without causing obvious visible injury. Unfortunately, it is extremely difficult to establish valid quantitative conclusions from this mass of literature because the results are so variable. This is because various parts of different species, cultivars or clones of plants are studied at different seasons, in different environmental conditions (soil, temperature, humidity, etc.) with a wide variety of exposures to pollutants using open-top chambers, fumigated field plots, controlled environment cabinets, and wind tunnels. Study of this literature reveals clearly the fact that SO_2 may reduce growth and yield in the absence of visible symptoms.

Invisible injury associated with relative changes in growth, yield, etc., can only be evaluated by having a valid control alongside for comparison. In the past, some have argued that a valid control is one in which plants are grown in atmospheres containing a 'minimum' of pollutant (i.e. charcoal-filtered air). The problem with this is that this situation is almost as unrealistic as some of those long-term exposures to continuous high levels of SO_2 which have actually only occurred momentarily as peak values in the outside environment. Comparisons of plant growth in charcoal-filtered air to those exposed to low levels of SO_2 are remarkable. Growth differences of as much as 30% are often measured, but 30% of nothing is still nothing. In many instances, the charcoal-filtered situation is an idealized situation and the low level treatment may represent a more realistic control upon which comparisons are to be made. Comparative work using charcoal-filtered controls can therefore only be done in areas known to have little SO_2 pollution, otherwise results from charcoal-filtered experiments should be treated with some caution. Furthermore, filters such as activated charcoal also remove NO_2, O_3, PAN and hydrocarbons (but not NO) and, consequently, hidden interactions between certain pollutants are unwittingly exposed. In short, these two types of experiment are measuring quite different differences but both have relevance. Perhaps comparative experiments should have two controls – one 'clean-air' and one 'charcoal-filtered'?

From studies so far, it has been concluded that the SO_2 concentrations prevailing in most agricultural regions of Europe, north America and elsewhere are generally not high enough to reduce significantly the yield of the major cereals. However, concentrations of 60 nl l^{-1} SO_2 are capable of causing a 7.5% reduction in one of the commonest grasses found in permanent pasture (perennial ryegrass, *Lolium perenne* L.) and, according to an OECD cost–benefit analysis carried out in 1981, this may be as high as a 25% reduction in growth. Yield reductions in grasses due to SO_2 are related to dosage (concentration multiplied by time) but are complicated by the additional presence of other pollutants, especially NO_2, which cause more-than-additive (synergistic) reductions in growth (see Chapter 11).

A wide variety of environmental conditions affect the growth of grasses and trees. The rate of air movement over a sward or canopy influences sensitivity to SO_2. If the wind speed is high, the boundary layer resistance is reduced significantly as the layer of still air directly above the leaves is ripped away. Seasonal factors also have to be taken into account. In grasses, inhibitory effects of SO_2 are greater (by as much as 50% in the late winter) under conditions of slow growth caused by poor light or low temperatures.

All types of environmental stress, not just those caused by air pollutants, affect growth and this may again change in the presence of SO_2. Figure 2.11 is a conceptual model which emphasizes that pollution injury cannot be divorced from other consequences of stress – each one is

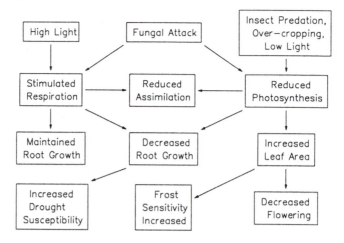

Fig. 2.11. Interrelationships between different types of plant stress.

interrelated. Moreover, age and spacing between plants are extra factors. Consequently, it is not surprising that divergent and apparently contradictory effects of SO_2 on plants have been reported in the past.

Health effects

Irritant properties

Sulphur dioxide and its derivatives differ markedly from the nitrogen oxides (but less so from O_3 and photochemical oxidants) in that they produce strong irritation upon the eyes and also within the nasal passageways. However, SO_3 on a molar basis produces more than four times the irritant response than SO_2 because it combines immediately with water to form sulphuric acid (Reaction 2.1). Sulphate in particulates can also cause similar irritation and the intensity of the effect is dependent upon the concentrations of the particulates and their sizes. Particles smaller than 1 μm are the most irritating but larger particles, as well as high concentrations of gaseous SO_2, also induce an involuntary coughing reflex. The eye irritation combined with this choking cough immediately draws the attention of those affected to the hazards of the surrounding atmosphere, although there is a wide variation in response susceptibility between individuals. SO_2 can also be detected by smell by certain individuals rather better than others, but at very high concentrations ($> 3 \, \mu l \, l^{-1}$) this sense of smell is quickly paralysed in all.

Figure 2.12 shows the major air passageways into the lungs which

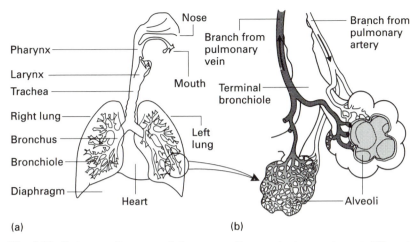

Fig. 2.12. Cut-away diagrams of the mammalian respiratory system at different scales of resolution.

terminate in the alveoli (fluid-lined air sacs) across which O_2 and CO_2 exchanges take place with the bloodstream. The mode of entry of SO_2 into this system, however, differs from those associated with nitrogen oxides, O_3 and photochemical oxidants because more than 95% of inhaled SO_2 is absorbed in the airways above the larynx (voice box) and less than 1% gets down into the lung alveoli when an individual is resting. Exercise increases the exposure of lung alveoli to SO_2 but, at the same time, the upper airways are constricted and the mucus lining dries up. This bronchoconstriction and dryness may induce chronic bronchitis in some individuals, especially those predisposed to asthma. The same group often also show a sensitivity to sodium sulphite used commercially as a sterilizing agent or as a preservative.

During severe short-term exposures to SO_2, sulphate and sulphite anions formed on the cell surfaces of the nasal linings penetrate the surface mucosal cells and bind to granules within mast cells beneath. This causes local release of histamine which then acts as a local modulator to cause constriction of the airways and initiates local inflammation. Ciliary cells lining parts of the airways often release extra mucus to carry away some of the harmful anions as the nose is blown or one coughs. Similarly, tears running from the eyes provide relief from the local irritation.

Although SO_2 is the oldest recognized harmful air pollutant affecting humans, recently other priorities (e.g. photochemical oxidant damage) have tended to disguise the harmful consequences to human health of this acidic gas. Because animals lack the coughing reflex, it has always been difficult to extrapolate results obtained from animals to likely

Table 2.3 Human response to different levels of SO_2 for different periods

Concentration ($\mu l\, l^{-1}$)	Period	Effect
0.03–0.5	Continuous	Condition of bronchitic patients worsened
0.3–1	20 seconds	Brain activity changed
0.5–1.4	1 minute	Odour perceived
0.3–1.5	15 minutes	Increased eye sensitivity[a]
1–5	30 minutes	Increased lung airway resistance, sense of smell lost
1.6–5	> 6 hours	Constriction of nasal and lung passageways
5–20	> 6 hours	Lung damage reversible if exposure ceases
20 upwards	> 6 hours	Waterlogging of lung passageways and tissues, eventually leading to paralysis and/or death

[a] Generally lower if aerosols, particulates or other pollutants are also present.

effects on humans. This lack of correlation, however, has not stopped a large number of animal studies over a long period (nearly 50 years) to elucidate mechanisms of action and to evaluate total risk even if the comparative risk to human health cannot be quantified accurately.

The US TLV for the five 8 h periods of a working week set at 5 $\mu l\, l^{-1}$ SO_2 appears to offer very little, if any, margin of safety. This inconsistency is due to the fact that the studies which were largely used to establish this limit showed that concentrations of 50 times ambient levels were said to produce 'little' distress in experimental animals. Consideration of Table 2.3 will show that levels of SO_2 well below 5 $\mu l\, l^{-1}$ have been shown to cause adverse human responses. It would appear that only the healthiest non-smoking adults are covered by this TLV and no margin of safety is allowed for interactions caused by other air pollutants, aerosols or particulates.

Industrial and urban hazards

Workers in certain industries (e.g. smelting, sulphuric acid production, etc.), who have been exposed longer and more regularly to SO_2 than others, are often claimed to have acquired a degree of tolerance to this gas. On the other hand, this group also accounts for most of the industrial fatalities and acute injuries. Sudden deaths resemble those due to asphyxiation but, as is more normal when death is delayed, subsequent post-mortem examination reveals oedema (enhanced water uptake into tissues), destruction of the ciliated covering of the air passages and, more significantly, invasion of the lungs by bacteria.

Table 2.4   Summary of the important epidemiological surveys into the relationship between long-term health and SO_2 plus particulates (from Economic Commission for Europe, 1984)

Country	*Annual average pollutant levels*		Effects noted
	SO_2 ($nl\,l^{-1}$)	*Particulates* ($\mu g\,m^{-3}$)	
USA	9.5	135	Higher incidences of acute respiratory diseases
USA	21	180	Increased rate of respiratory symptoms; decreased lung function
UK	38	200	Elevated incidence rates of respiratory problems especially in children
Poland	48	270	More chronic bronchitis and asthmatic disease in smokers
Russia	48	285	Enhanced symptoms of respiratory disease
UK	86	360	Higher frequency of respiratory symptoms; decreased lung function in children

During accidental non-fatal exposures, victims experience inflammation of the eyes, nausea, vomiting, abdominal pain, a sore throat, bronchitis, and often pneumonia. Normally, the lungs are (and have to be) highly sterile but this weakening of natural antibacterial mechanisms points the way to emergency treatment of SO_2-exposure cases. Prompt use of oxygen and bronchodilators along with heavy doses of antibiotics prevents permanent damage and may save life.

Mercifully, sudden accidental acute cases are rare, although the possibilities of chronic poisoning by SO_2 over longer periods at low levels are greater for the general population. The literature which deals with and confirms the correlation between chronic chest disease and levels of SO_2 in urban air is vast, but the evidence suggesting that chronic exposure to SO_2 and particulate matter plays a part in the cause and development of chronic respiratory disease has only been established with difficulty. The problem resides in the multitude of other possible causes which often cannot be fully excluded during large-scale epidemiological studies. Adequate allowance must be made for previous medical history, tobacco smoking, age, sex, family size, racial origin, socio-economic, seasonal and meteorological variables.

Nevertheless, high levels of SO_2 and particulates do give rise to shorter lives and poorer health, according to epidemiological studies which have been carried out in the USA, the UK, and elsewhere (Table

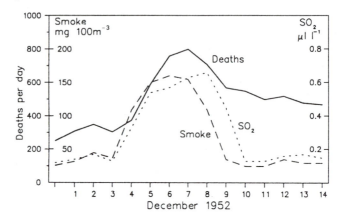

Fig. 2.13 The increased mortality of Londoners in December 1952 has been ascribed to the smog-like conditions prevailing during the first part of this month which contained high levels of SO_2 and smoke particles.

2.4). The UK studies have also indicated higher rates of lung cancer in affected groups. The classic increase in death rate of Londoners during the 'smog' of 1952 is often quoted (Fig. 2.13). But high incidences of death (mortality studies) and peaks in air pollution are often unrelated. The highest average level of SO_2 during the London smog of 1952 was actually 0.7 $\mu l\ l^{-1}$ – well inside the US TLV. The main findings of the major epidemiological studies are shown in Table 2.4 alongside the annual pollutant levels at which the main symptoms were noted. It is studies like these that caused the WHO to recommend a guideline of 15–23 $nl\ l^{-1}$ (40–60 $\mu g\ m^{-3}$) as an annual mean which is thought to incorporate a safety factor of between 2 and 3.

Toxicity of sulphur dioxide and sulphite within tissues

Most human cell tissues have more than adequate amounts of sulphite oxidase activity (Reaction 2.21) to remove bisulphite or sulphite almost as soon as it appears. The average dietary intake of bisulphite from preserved foods is less than 0.2 mmol d^{-1} – much less than that caused by internal breakdown of S-containing amino acids (about 25 mmol d^{-1} person^{-1}). The liver alone appears to have more than 200 times the capacity needed to oxidize these amounts. It has been calculated that a person breathing 5$\mu l\ l^{-1}$ SO_2 for 8 h would take up the equivalent of 1.3 mmol of bisulphite and sulphite during the day. Apparently, the lungs have more than 100 times as much sulphite oxidase as is necessary to remove this, but such calculations of capacities are spread over a whole day and throughout the whole lung. It is far from certain that these intense local accumulations in the epithelial tissues of the lungs or nasal regions can be removed immediately before adverse localized damage occurs. The major mechanisms involved are those of free radical and nucleophilic attack similar to those described earlier.

Studies with animals exposed to radioactive $^{35}SO_2$ have also shown that lung sulphite oxidase activities are insufficient to prevent other reactions in the lungs and from ^{35}S-labelled bisulphite entering the blood. Labelled bisulphite appears to be immediately bound up as ^{35}S-sulphocysteine (Reaction 2.22) which is then taken into either the liver or other tissues. Because the liver has all the necessary detoxifying enzymes, it quickly returns radioactivity back to the blood as ^{35}S-labelled sulphate which is then quickly eliminated by the kidneys. The remainder trapped in the other tissues slowly returns to the blood and eventually emerges in the urine.

The question that then remains is 'How do results from animal studies carried out at high levels of SO_2 relate to humans exposed to lower amounts?' There seems to be general agreement that any mutagenetic consequence due to cytidine alterations (see earlier) are unlikely but enhanced levels of sulphocysteine in nerve tissues may be cause for concern. Low levels of sulphocysteine given intravenously to animals have, for example, been shown to cause brain abnormalities. It is adverse reactions within lung tissue and close to the airways that have most relevance to humans. Epidemiological studies showing respiratory problems to be the main consequence of exposure to SO_2 confirm this statement.

Unfortunately, few studies to understand the mechanism of these problems have been undertaken. Tissue cultures of alveolar macrophages (phagocytic cells associated with removal of debris) and lymphocytes (non-phagocytic cells associated with antibody production) have been isolated from both animal and human lungs. Initially, when such cells are exposed to SO_2, sulphite or bisulphite, they show elevated levels of a plasma membrane-bound enzyme called ATPase. They also have depleted levels of ATP which implies a general reduction in cell capabilities. External levels of lysozyme (a general protein-degrading enzyme which is secreted by macrophages) are also increased by SO_2. This appears to act as a trigger for fresh synthesis of lysozyme to take place inside macrophages. Moreover, such protein breakdowns in lining tissues have undesirable effects upon the integrity of the surrounding aveolar cells and this diverts the attention of the macrophages from any ongoing bacterial infection.

As well as changes in levels of enzymes, the properties of the surrounding cell membranes may also be altered by SO_2, bisulphite or sulphite which may mean that lymphocytes are not able to recognize and bind antigens. Consequently, they do not produce the appropriate antibodies or recognize (and respond to) foreign or aberrant cells. This reduced immune surveillance then gives rise to the bronchitic and respiratory problems identified by the epidemiological surveys.

Similar membrane changes occur in the macrophage cells of lungs. Normally, lymphocytes produce a lymphokine referred to as MIF (macrophage inhibitory factor) which controls the movement of macrophages. After exposure to SO_2, macrophages lose their ability to respond to MIF and show increased migratory tendencies, thus avoiding the phagocytic (bacterial engulfing) task identified by the lymphocytes.

It is events like these that give rise to the long-term respiratory problems of humans caused by atmospheric SO_2.

Further reading

Anderson, J.W. *Sulphur in Biology*. Studies in Biology No. 101, Edward Arnold, 1978, London.

Brimblecome, P. *Air Composition and Chemistry*. Cambridge University Press, 1986, Cambridge.

Economic Commission for Europe (OECD). *Air-borne Sulphur Pollution*. Air Pollution Study No. 1, United Nations, 1984, New York.

Mansfield, T.A. (ed.) *Effects of Air Pollutants on Plants*. SEB Seminar Series No. 1, Cambridge University Press, 1976, Cambridge.

Muth, O.H. (ed.) *Sulphur in Nutrition*. AVI, 1970, Westport, Connecticut.

Nriagu, J.O. (ed.) *Sulphur in the Environment*. John Wiley & Sons, 1978, New York.

Saltzman, E.S. and Cooper, W.J. (eds) *Biogenic Sulfur in the Environment*. Amer. Chem. Soc. Symp. No. 393, 1989, Washington, DC.

Stern, A.C. *Air Pollution* (3rd edn), vol. 2, *The Effects of Air Pollution*. Academic Press, 1977, New York.

Unsworth, M.H. and Ormrod, D.P. (eds) *Effects of Gaseous Air Pollution in Agriculture and Horticulture*. Butterworths, 1982, London.

Waldbott, G.L. *Health Effects of Environmental Pollution* (2nd edn). C.V. Mosby, 1978, St Louis, Missouri.

Winner, W.E., Mooney, H.A. and Goldstein, R.A. (eds) *Sulphur Dioxide and Vegetation: Physiology, Ecology and Policy Issues*. Stanford University Press, 1985, Stanford, California.

Chapter 3

NITROGEN OXIDES

'It is a fire, it is a coale – whose flame creeps in at every hole'
from 'The Hunting of Cupid' by George Peele [1558–1597].

Formation and sources

The nitrogen cycle

Nitrogen dioxide (NO_2) and nitric oxide (NO) are not the predominant nitrogen oxides of the atmosphere but they are the ones which appear to give most problems in the troposphere. Air chemists often use the abbreviation NO_x for these two pollutants but, unfortunately, the use of the term NO_x implies that the biological effects of NO_2 are similar to those of NO, a component sometimes in the largest proportion. More seriously, it is never clear if NO_x includes or excludes the greenhouse gas, nitrous oxide (N_2O; dinitrogen monoxide), which is also implicated in stratospheric O_3 depletion, but it is more than 10 times as prevalent in the atmosphere as NO_2. If one uses the term 'nitrogen oxides' the meaning is clear. It includes NO_2 and NO, as well as N_2O, N_2O_3, N_2O_5, etc. – each of which has different effects on living systems. It is better if the abbreviation NO_x is not used for biological purposes because it induces a belief in similar effects being caused by different nitrogen oxides – which is *not* the case.

Not all nitrogen (N)-containing air pollutants are oxidized. Ammonia (NH_3), for example, is released into the atmosphere as a result of the decay of waste products from animals or as escaping gas during the manufacture of artificial fertilizers. As NH_3 or its ionic product ammonium (NH_4^+), these reduced forms of N in the atmosphere are especially harmful to ecosystems sensitive to excessive inputs of N. Releases of reduced N compounds to the atmosphere and their effects are covered separately in Chapter 4.

As N circulates between plants and animals and microbes in the soil, wide differences exist between different reduced and oxidized components. The electronic configuration about a N atom may vary widely, which permits a number of different oxidation–reduction states of N to exist (Fig. 3.1). This, in turn, gives rise to a wide range of N-based

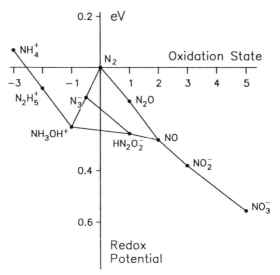

Fig. 3.1. Oxidation states of nitrogen at pH 7.0 in relation to redox potentials. As electronegativity represents a higher energetic state than electropositivity, electrons flow down such a gradient.

compounds and together these features dominate the biological nitrogen cycle (Fig. 3.2). Homogenic impacts upon global exchange of N are considerable (Table 3.1) but the bulk of the global N cycling remains microbial.

Nitrous oxide

The most abundant nitrogen oxide in the atmosphere is N_2O. This is mainly released by denitrification (Fig. 3.2) which is an undesirable agricultural process carried out by certain soil microbes using nitrate instead of O_2 to respire. Soil conditions where O_2 levels are low favour denitrification. Consequently, stagnant, waterlogged or compacted soils are major sources of N_2O, and some of the main purposes of ploughing and draining are to discourage this anaerobic process.

Other soil microbes also carry out nitrification in which the conversion of ammonium or nitrite to nitrate by O_2-dependent microbes is encouraged by aeration. One of the basic objectives of arable farming is to maximize microbial nitrification and, at the same time, encourage fixation of atmospheric N_2 into NH_3 by the enzyme nitrogenase (Fig. 3.2). By applying artificial fertilizers, however, a farmer deliberately disturbs the natural balance between these different processes and discourages microbial N_2 fixation. The alternative is to plant legumes or clovers in rotation but prevailing economics rather than science are

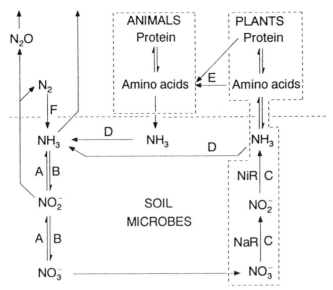

Fig. 3.2. The nitrogen cycle as it relates to plants, animals, the soil, and the atmosphere. The various processes are as follows: **A** = denitrification or dissimilatory nitrogen metabolism which releases N_2O; **B** = microbial nitrification; **C** = assimilatory nitrogen metabolism; **NaR** = nitrate reductase; **NiR** = nitrite reductase; **D** = decay after death or putrefaction; **E** = digestion; and **F** = nitrogen fixation.

Table 3.1 Global transfer rates of nitrogen by various processes[a]

Process	$Tg\ N\ a^{-1}$
Denitrification/nitrification	160
Natural nitrogen fixation	150
Fires and combustion	70
Industrial fixation	40
Ionic exchange by rain, etc.	80
Run-off to oceans	35

[a] Estimates of the forms and global amounts of nitrogen compounds vary but are very similar to the global rates of transfer because, with the exception of N_2O (which has an atmospheric lifetime of 150 years), most of the others have lifetimes measured in weeks. Because of this, the global reservoir of N_2O is by far the largest (1500 Tg N) followed by NH_3 (1.6 Tg N), ammonium (0.5 Tg N), NO_2 (0.4 Tg N), NO (0.2 Tg N), nitric acid (0.2 Tg N) and nitrate (0.1 Tg N).

the deciding factors. Any excess artificial fertilizer remaining after application, which has not drained away, is usually removed by denitrification. Increasing global use of artificial fertilizers therefore increases atmospheric levels of N_2O because denitrification is encouraged.

Global production of N_2O by denitrification each year amounts to about 5% of the total global pool of atmospheric N_2O but, because N_2O is a relatively unreactive molecule, it has a long residence time in the atmosphere of 20 years or more. Any subsequent atmospheric increase in N_2O levels due to enhanced denitrification will therefore develop slowly but persist for a long time.

Small amounts of N_2O are also produced by the process of nitrification (Fig. 3.2) as a result of incomplete conversion of ammonium to nitrite. Consequently, small losses of fixed nitrogen occur at different stages during nitrification. In compensation, N_2 fixation involves some consumption of N_2O as it is an alternative substrate for nitrogenase which is found in all organisms capable of fixing N_2.

The big unknown involves the oceans of the world. Their relative alkalinity as compared to those of soils favours denitrification, and hence N_2O production, because the energy yield of the denitrification reaction is inversely related to pH. This is the case especially in deep waters where O_2 is scarce or in estuaries where O_2 may be depleted. Sewerage discharges which cause a general decline in O_2 content, causing rivers to become eutrophic, therefore encourage denitrification.

Lack of removal mechanisms for N_2O in the troposphere means that N_2O molecules drift slowly upwards into the stratosphere where they may undergo either photolysis or reaction with atomic oxygen. Photolysis (Reaction 3.1) produces molecular N_2 and O_2 which has no major implications, but Reaction 3.2 which forms NO has a strong catalytic influence upon reducing O_3 levels in the stratosphere (see Chapter 7). The links between increasing global use of nitrate fertilizers, enhanced denitrification, and possible influences upon stratospheric O_3 have been the cause for some concern. Unfortunately, because this process involves a long time period, it will be impossible to reverse if depletions in stratospheric O_3 due to this cause are ever detected (see Chapter 7). Furthermore, N_2O is an important greenhouse gas. The implications of this are covered in Chapter 8. Discouraging the use of nitrate-based fertilizers is bound to be unpopular, especially in developing countries. Genetic transfer of the genes associated with N_2 fixation or N uptake to important economic plants such as cereals and grasses may yet be able to retrieve this situation.

$$2N_2O \overset{\text{light}}{\Rightarrow} 2N_2 + O_2 \qquad (3.1)$$

$$N_2O + O \Rightarrow 2NO \qquad (3.2)$$

Combustion

NO is mainly formed by the combination of atmospheric N_2 and O_2 at high temperatures with a lesser contribution coming from N-containing components in the fuel. Reaction 3.3 is often quoted as the major process, but actually it takes place in separate reactions (Reactions 3.4 and 3.5) involving highly reactive atoms created within a flame – the second being much faster than the first. In the fuel-rich regions of a flame, highly reactive hydroxyl (OH$\cdot$) radicals (see Appendix 1) form NO (Reaction 3.6) but HCN, NH_3, or amines ($-NH_2$, $=NH$ or $\equiv N$) have also been identified as precursors of NO.

$$N_2 + O_2 \Rightarrow 2NO \tag{3.3}$$

$$N_2 + O \Rightarrow NO + N \tag{3.4}$$

$$N + O_2 \Rightarrow NO + O \tag{3.5}$$

$$N + \cdot OH \Rightarrow NO + H \tag{3.6}$$

Odd as it may first appear, the amount of NO produced decreases as the N content of the fuel rises. By contrast, removal of S from fuels increases the amount of NO produced. Therefore, burning of heavy oil, which usually has a low N content, leads to NO emissions as great as those from power plants using coal with much higher N content.

Much can be done to reduce NO emissions from furnaces in terms of the fuel used, the design of burners, and the operating conditions. For example, the particle or droplet sizes of the fuel may be reduced which then provides larger surface areas to encourage fuller combustion (Reactions 3.7–3.9). Other parameters which may be adjusted are the temperatures, pressures, oxygen availability, mixing rates, and residence times. Indeed, it is several times cheaper to redesign burners in order to reduce emissions of nitrogen oxides (and hence acidity) than to retrofit desulphurization equipment for the reduction of total acid-forming emissions. Another approach, used in Japanese power plants, has been to inject NH_3 into the combustion process by a process known as selective catalytic reduction. However, this process (Reaction 3.10) is only efficient at temperatures below 1000°C – above this the NH_3 tends to form NO once again.

$$C + 2NO \Rightarrow CO_2 + N_2 \tag{3.7}$$

$$2C + 2NO \Rightarrow 2CO + N_2 \tag{3.8}$$

$$2CO + 2NO \Rightarrow 2CO_2 + N_2 \tag{3.9}$$

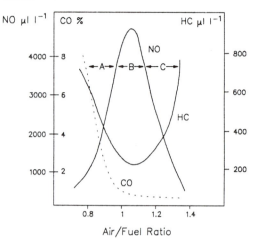

Fig. 3.3. Emissions of carbon monoxide (CO), nitric oxide (NO) and unburnt hydrocarbons (HC) from internal combustion engines operated at different air/fuel ratios. **A** = range of mixtures used in the past; **B** = present-day engines; **C** = 'lean-burn' engines.

$$4NO + 4NH_3 + O_2 \Rightarrow 4N_2 + 6H_2O \qquad (3.10)$$

In most developed countries, road transport contributes about 30% of the total emissions of nitrogen oxides as opposed to 45% from power plants and 25% from domestic and general industrial sources. In many countries, improvements to fuel combustion by redesigning vehicle engines and fitting catalytic exhaust systems as standard to reduce emissions of NO have been recent developments. Alterations to combustion within engines go hand in hand with improved catalytic conversion of exhaust gases. Increasing the ratio of air to fuel used by engines also reduces greenhouse gas emissions. Justification for this 'lean-burn' concept is illustrated by the gaseous changes in Fig. 3.3. The problem that still remains for engine designers is to reduce the emissions of unburnt hydrocarbons (HCs) as leaner air : fuel mixtures are used.

Expensive three-way platinum/rhodium catalysts are sometimes incorporated into vehicle exhaust systems to encourage all three removal reactions (Reactions 3.11–3.13) rather than the cheaper oxidation catalysts often specified. The latter only carry out Reactions 3.11 and 3.12 but have little effect on overall nitrogen oxide emissions. The problem with three-way catalysts is that removal of unburnt HCs and CO requires plenty of oxidant but removal of NO is hindered by excess oxidant (see Fig. 3.4). Only a narrow window of oxidant availability exists for all three conversions to occur and it is difficult to maintain them all throughout different driving conditions and the lifetime of the catalytic exhaust system.

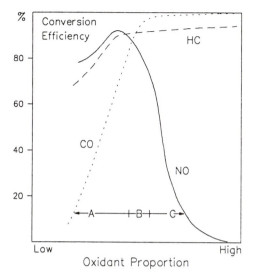

Fig. 3.4. Conversion efficiencies (%) of carbon monoxide (CO), nitric oxide (NO) and unburnt hydrocarbons (HC) by catalysts included in car engine exhaust systems. The effective regions of either reduction (**A**) or oxidation (**C**) catalysts of single or dual-bed systems operate either side of region **B** which is the effective range of platinum/rhodium three-way catalytic systems.

$$2CO + O_2 \Rightarrow 2CO_2 \qquad\qquad (3.11)$$

$$(CH_2) + O_2(+2NO) \Rightarrow CO_2 + 2H_2O\,(+N_2) \qquad (3.12)$$

$$2CO + 2NO \Rightarrow 2CO_2 + N_2 \qquad\qquad (3.13)$$

All systems involving lean-burn engines or three-way catalytic exhaust systems also require much greater control over the fuel input to the engine over the wide range of driving conditions. Computerized fuel injection with feedback loops from exhaust sensors will therefore become standard and increase costs, but again this will be complemented by overall fuel savings and reductions in greenhouse gas emissions. Legislation favouring exhaust improvements also has an additional benefit because it hastens a general reduction in the lead content of fuels and, hence, lead emissions to the atmosphere, because lead quickly poisons exhaust catalysts.

Other industrial processes also produce significant emissions of nitrogen oxides. Ammonium nitrate is widely used as a nitrogenous fertilizer for agriculture and is produced in large quantities. It is usually synthesized from the reaction of NH_3 with nitric acid (Reaction 3.14) – the NH_3 required being generated by the Haber process (Reaction 3.15) while nitric acid is formed from the catalytic oxidation of NH_3 (Reactions 3.16–3.18). Escapes and emissions of nitrogen oxides, NH_3, and particulate ammonium nitrate to the atmosphere from such factories are significant and often give rise to considerable local pollution.

$$HNO_3 + NH_3 \Rightarrow NH_4NO_3 \tag{3.14}$$

$$N_2 + 3H_2 \Rightarrow 2NH_3 \tag{3.15}$$

$$4NH_3 + 5O_2 \Rightarrow 4NO + 6H_2O \tag{3.16}$$

$$2NO + O_2 \Rightarrow 2NO_2 \tag{3.17}$$

$$3NO_2 + H_2O \Rightarrow 2HNO_3 + NO \tag{3.18}$$

Atmospheric oxidations

Once in the atmosphere, NO is rapidly oxidized to NO_2 by reaction with O_3 (Reaction 3.19) which is counteracted by a photochemical back-conversion (Reaction 3.20) while the O_3 is regenerated by Reaction 3.21. The overall catalytic cycle (i.e. Reaction 3.19 working against Reactions 3.20 and 3.21) allows a photostationary state to be established which permits significant atmospheric concentrations of NO, NO_2 and O_3 to exist in the presence of each other. The proportions are controlled by the light flux, the temperature, and the presence of particulates and unburnt HCs (see Fig. 6.1).

$$O_3 + NO \Rightarrow NO_2 + O_2 \tag{3.19}$$

$$NO_2 + \text{light} \Rightarrow NO + O \tag{3.20}$$

$$O + O_2 + M^1 \Rightarrow O_3 + M \tag{3.21}$$

Apart from photolytic breakdown (Reaction 3.20), a variety of alternative reactions may be responsible for the removal of NO_2 (see Fig. 3.5). In bright light, photochemically generated ˙OH radicals or O_3 produce nitric acid (HNO_3) by means of Reaction 3.22 or Reactions 3.23–3.25 which return ˙OH radicals back into the system (Reaction 3.26). Other products stemming from the formation of highly reactive NO_3 include NO_2 and its equilibrium alternative, N_2O_4 (Reactions 3.27 and 3.28) or more NO by the action of light (Reaction 3.29). Reactions 3.23 and

[1]See Reactions 2.2 and 2.7.

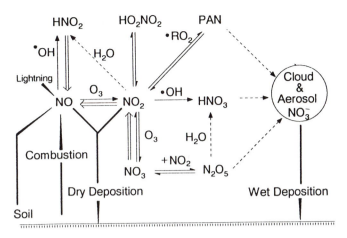

Fig. 3.5. Atmospheric interconversions of nitrogen oxides (courtesy of Drs Cox and Penkett, AERE, Harwell, UK, and D. Reidel Publ. Co., Dordrecht, The Netherlands). The open arrows indicate light-dependent reactions. (See also Fig. 6.1 for a simplified version.)

3.27–3.29, therefore, assist Reactions 3.19 and 3.20 in maintaining the major equilibria between NO and NO_2 in the atmosphere. Current estimates predict that each NO radical destroys on average 300 O_3 molecules (Reaction 3.19) before it is trapped.

$$\cdot OH + NO_2 + M^1 \Rightarrow HNO_3 + M \qquad (3.22)$$

$$NO_2 + O_3 \Rightarrow NO_3 + O_2 \qquad (3.23)$$

$$NO_3 + NO_2 + M^1 \Rightarrow N_2O_5 + M \qquad (3.24)$$

$$N_2O_5 + H_2O \Leftrightarrow 2HNO_3 \qquad (3.25)$$

$$HNO_3 + light \Rightarrow \cdot OH + NO_2 \qquad (3.26)$$

$$NO_3 + NO \Rightarrow 2NO_2 \Leftrightarrow N_2O_4 \qquad (3.27)$$

$$NO_3 + NO \Rightarrow 2NO_2 \qquad (3.28)$$

$$NO_3 + light \Rightarrow NO + O_2 \qquad (3.29)$$

Other important atmospheric reactions involve the formation of both peroxyacyl nitrates (PANs; Reaction 3.30) and peroxynitric acid (HO_2NO_2). Most PANs have a half-life of about 1 h but peroxynitric acid is more unstable and lasts only a few seconds in the lower troposphere. However, they are all thermally stable in the colder upper regions of the troposphere and lower stratosphere where interactions with ˙OH radicals are more prevalent. Formation of peroxynitric acid relies on the presence of the hydroperoxyl radical (HO_2˙) which is formed by the rapid addition of atomic hydrogen to O_2. This radical and the conditions for its formation (i.e. ionizing radiation) are again more likely in the stratosphere.

Nitrous acid (HNO_2) forms rapidly in the gas phase by reaction between NO, NO_2 and H_2O (Reaction 3.31) but decomposes to NO and nitric acid in aqueous solution by a series of interconversions (Reactions 3.32–3.34). This decomposition, however, is slowed in dilute solutions because Reaction 3.32, which forms NO_3, only occurs in concentrated solutions.

$$CH_3CO.O_2 + NO_2 + M \Rightarrow CH_3COO_2NO_2 + M \qquad (3.30)$$

$$NO + NO_2 + H_2O \Rightarrow 2HNO_2 \qquad (3.31)$$

$$2HNO_2 \Rightarrow N_2O_3 + H_2O \qquad (3.32)$$

$$2N_2O_3 \Rightarrow N_2O_4 + 2NO \qquad (3.33)$$

$$N_2O_4 + H_2O \Rightarrow HNO_2 + HNO_3 \qquad (3.34)$$

The composition of nitrogen oxides in the atmosphere is therefore very complicated. A multitude of different reactions (see Fig. 3.5) are influenced to a greater or lesser extent by temperature, light flux, and different levels of dilution because all these molecules, atoms or radicals have to collide in order to react. The chances of this are reduced as pressure falls with increasing altitude but, as molecules rise upwards, some of the intervening collisions with unreactive molecules are reduced and radiation levels increase. Consequently, half-lives of reacting species vary with altitude and some intermediates, which exist for very short periods at ground level, persist for longer in the stratosphere. On the other hand, those species which are relatively inert at ground level have a chance to be slowly transported upwards into the stratosphere where they subsequently react. N_2O is a good example of this possibility.

Dry deposition

All nitrogen oxides are removed from the atmosphere by absorption at ground level by land or ocean surfaces. As Chapter 1 has emphasized, the velocity of deposition of each atmospheric gas differs widely and depends upon the nature of the uptake surfaces. Rates for chemical species like nitric acid are very high because they are so reactive, while the deposition velocity of NO_2 is lower than that of SO_2 or O_3 but much higher than that of NO (Table 1.4) or N_2O. Estimates of global wet and dry deposition rates of both nitrate and nitric acid (which are difficult to separate) amount to about 25 Tg N a^{-1} while dry deposition of NO_2 accounts for an additional third of this amount.

Measurements of deposition rates are complicated because the gases may interact in many different ways. For example, undissolved nitric acid forms ammonium nitrate aerosol (Reaction 3.14) – a salt which has a high dissociation constant so that significant equilibrium levels of both NH_3 and nitric acid are present. The relative deposition rates of different nitrogen-containing compounds are as follows:

$$HNO_3 = NO_3^- > NH_3 > NH_4^+ > NO_2 > NO > N_2O$$

As may be expected, dry deposition rates of particles range over several orders of magnitude, depending upon the sizes of the particles (see Chapter 1) as well as the degree of turbulence in the atmosphere.

Plant effects

Leaf access

Movements of CO_2 and H_2O in and out of a leaf have been examined in great detail by plant scientists. Research so far shows that most pollutant gases like SO_2 and O_3 move into a leaf in a manner similar to that of CO_2 (see Fig. 2.8). Generally, the movement of the nitrogen oxides into leaves is subject to similar diffusive resistances as found for SO_2 but the stomata are not the only path of entry for nitrogen oxides. Cuticular resistances against NO_2 entry are lower than those for SO_2 or O_3. This means that, even when stomata are closed, some nitrogen oxides possibly enter a leaf through the epidermal layers even though this is a small fraction of the whole.

Inside a leaf, the exposed surface area of moist cell wall below is considerably larger than the external surface area of a leaf and this provides a large surface area for the absorption of nitrogen oxides. Solubility of nitrogen oxides in the extracellular fluid is therefore an important factor in determining their rate of uptake. Both gaseous NO_2 and NO are only slightly soluble in H_2O, but NO_2 vastly increases its apparent solubility by reacting with H_2O to produce nitric acid (Reaction 3.18). Gaseous NO, however, does undergo a comparatively slow reaction with H_2O to form both nitric acid and nitrous acid (HNO_2, Reactions 3.31–3.34). Normally, nitrous acid is oxidized in solution to nitric acid by O_3, H_2O_2 and other oxidizing agents, but it may be converted back to NO when reducing agents like ferrous iron are present. Measurements show that gaseous NO is taken up by plants about a third as well as NO_2.

Unfortunately, the natural world does not have a convenient radio-active isotope of nitrogen which could be used to track the pathway of nitrogen oxide uptake. The one that exists for longest is ^{13}N which has a half-life of only 10 minutes. Studies using O_2 are similarly hampered. In such cases, non-radioactive isotopes of N and O are used and the labelling is followed by differences in mass of the isotopes using a mass spectrometer. This is technically more difficult, less convenient and often not so discriminating. However, in experiments using both ^{15}NO gas and ^{18}O-labelled nitrate in solution, the ^{15}NO has been shown to move into both nitrite and nitrate and the ^{18}O into NO.

The cell wall bathed by the extracellular fluid is a very complicated matrix. A dense mass of cellulose fibrils provides a structural framework against which the turgid cells press and by doing so provide the structural rigidity of plant tissues. Within this *milieu*, ions, undissociated molecules and gases diffuse with differing degrees of ease towards and across the cell membrane beneath – yet another complex environment with both H_2O-attracting (hydrophilic) and H_2O-repelling (hydrophobic) phases. Some molecules are permitted to cross membranes passively, some are actively taken up by specific uptake, and others are not able to enter. In the case of the products of nitrogen oxides in solution in plants, their uptake appears to be a passive process.

As with SO_2, the uptake of nitrogen oxides into the extracellular fluid and their crossing of the cellulose wall, cell membrane, and cytoplasm to a site of action may be interpreted as a series of resistances collectively called the mesophyll resistance (see Fig. 2.8). In practice, this is difficult to measure, but some of the mechanisms which make a plant either tolerant or sensitive to the effect of a pollutant such as NO can only be ascribed to differing mesophyll resistances. Those plants that make it more difficult for the products of the nitrogen oxides in solution to enter, for example, are at an advantage over those that permit easy access. As shown later, there are mechanisms in a plant cell which induce the removal of the products of pollutants, especially those caused by nitrogen oxides.

Root uptake

Tracer experiments using $^{15}NO_2$ have shown that direct incorporation through the leaves (and stems) and uptake by roots after NO_2 has been absorbed into the soil are competing processes. However, amounts of $^{15}NO_2$ taken up by roots via the soil are small compared to direct incorporation through the leaves. Consequently, the soil route is only important during long-term exposures to NO_2. Nevertheless, investigations have shown that this indirect route via the roots does involve substantial inputs of nitrogen derived from atmospheric NO_2.

Even for soils that are intensively cultivated and have heavy additions of artificial or natural fertilizers, the additional input of N from the atmosphere is not negligible. Recommended applications of N-based fertilizer to grass swards should not exceed 200 kg N ha^{-1} a^{-1} because, at this level, there is a danger of exceeding nitrate run-off levels of the CEC directive set at >50 mg nitrate l^{-1}. As a result, levels of N application are often lower than this (*ca.* 100 kg N ha^{-1} a^{-1}). Around conurbations, inputs of N by wet and dry deposition from the atmosphere are of the same order of magnitude. Atmospheric inputs in excess of 40 kg N ha^{-1} a^{-1} have been recorded 50 km from London and much of this arrives in the winter, so only about half is taken up by a crop of winter wheat, for example, and over 30% is lost as run-off.

Similarly, for forests, scrub or tundra which are naturally N-limited, the extra N entering these ecosystems from deposition of nitrogen oxides becomes a major factor especially when N for extra growth is scarce and often unavailable. For some ecosystems, this extra N causes great disruption. Certain mires, for example, which are adapted to low N are invaded by alien species which normally need extra nitrogen to flourish.

Environmental conditions around roots also influence response of plants to NO_2 but factors such as H_2O availability, soil temperature, and mineral nutrient supply can affect the sensitivity of plants to various pollutants. At low levels of soil nitrogen, plants exposed to low concentrations of nitrogen oxides often show stimulated growth. For example, certain plants (e.g. grasses) grown on nitrogen-deficient soils and exposed to nitrogen oxides show a slight improvement in yield, whereas plants from the same stock but grown with an adequate supply of N fertilizer have decreased yield and sometimes show evidence of visible damage. Similar results have been found with other pollutants such as SO_2, O_3, and NH_3 in the presence or absence of other nutrients like sulphur, phosphorus and potassium.

Supplies of N to the leaves from the roots also influence response of plants to atmospheric NO_2. Root:shoot ratios decrease with increasing N availability, giving increased 'leafiness' and poorer root development. Consequently, increased availability of N to roots increases the amount of leaf area available for pollutant uptake relative to the whole plant biomass.

In most species, capacities for fixing CO_2 and transpiring H_2O are both directly related to the N content of a leaf – features which also have implications for vegetational response to climate change (see Chapter 8). Consequently, any disturbance caused by nitrogen oxides, either by acting as alternative sources of N or by affecting natural processes of C:N allocation, will have important consequences.

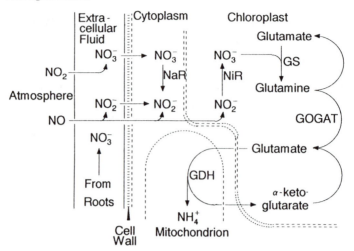

Fig. 3.6. Uptake and metabolic pathways of nitrogen oxides into plant tissues from the atmosphere, through the extracellular fluid, cell walls, cell membranes and cytoplasm into chloroplasts. Enzymes involved include nitrate reductase (NaR), nitrite reductase (NiR), glutamine synthetase (GS), glutamate synthase (GOGAT) and glutamate dehydrogenase (GDH).

Damage or benefit?

Plate 2 shows the effect of NO on two cultivars of lettuce grown hydroponically under conditions of CO_2 enrichment (1000 µl l^{-1}, typical horticultural glasshouse conditions). The levels of NO are admittedly high (2 µl l^{-1}) for most outdoor locations but in commercial glasshouses, or even indoor living conditions, they are quite common (see later). There are several major points arising from Plate 2. Generally, nitrogen oxides reduce rather than enhance growth but rarely cause visible injury. Plants also fail to take advantage of the extra N in nitrogen oxides even when they are N-limited (front and third rows growing at 0.5 mM nitrate have minimal supplies of N). Finally, there are wide variations in response to both nitrogen oxides and N availability between species and even between cultivars.

   The reduction of nitrite and the incorporation of NH_4^+ into glutamate (Fig. 3.6; see also Fig. 2.10) both require energy supplied from photosynthesis. Consequently, any increase in the flux of N through this pathway inevitably withdraws both energy and fixed C away from other reactions. However, it is unlikely that simple competition can explain the inhibitory effects of nitrogen oxides on photosynthesis and growth. Indeed, increasing CO_2 concentrations to grow horticultural crops often fails to produce bigger plants when they are also polluted with nitrogen oxides.

Levels of nitrite are known to rise in plants exposed to NO and NO_2 at lower concentrations and there may also be increases in nitrous acid, NH_3, and NH_4^+ ions as well. Nitrous acid reacts to form N- or C-containing nitroso-derivatives (Reaction 3.35) but when diluted the slowness of Reaction 3.32 and the decreased likelihood of the NO^+ species being produced at neutral pH make this unlikely. However, attack of nitrous acid on amino acids is a possibility (Reaction 3.36) but how important such a process is in biological systems has not been ascertained.

$$\begin{array}{c} R \\ \diagdown \\ NH + HNO_2 \Rightarrow \\ \diagup \\ R^1 \end{array} \begin{array}{c} R \\ \diagdown \\ N-N=O + H_2O + N_2 \\ \diagup \\ R^1 \end{array} \qquad (3.35)$$

$$RNH_2 + HNO_2 \Rightarrow ROH + H_2O + N_2 \qquad (3.36)$$

NO_2 will attack lipids to isomerize carbon:carbon double bonds from *cis* to *trans* forms or cause peroxidation (see Chapter 6). However, long-term exposure of plants to high atmospheric concentrations of NO_2 has failed to show any evidence of such reactions. By contrast, NO is a free radical ($\cdot N=O$) and reacts with metal atoms such as iron and copper even if these are enzyme bound. As these reactions often involve regulatory processes, it is now believed that this is the main reason why nitrogen oxides (especially NO) inhibit plant growth.

Reports of growth reductions caused by low concentrations of nitrogen oxides vary widely between species. Some grasses show no effects of fumigation (0.06 μl l^{-1} for 20 weeks) while others have more than 55% reduction in dry weight. The amount of damage varies in severity according to concentration and length of exposure, as well as with plant age, light flux, humidity, temperature, and season (see previous sections). There are even a few reports of beneficial effects of nitrogen oxides upon plant growth but these are confined to a few species.

Symptoms are often divided into 'invisible' and 'visible' injury (see Chapter 1). Following exposure to typical atmospheric levels of nitrogen oxides, instances of visible injury are very rare and can be confused with visible damage caused by SO_2. Characteristic chlorotic areas on the leaves associated with necrotic areas have been noted on a number of species exposed to high concentrations of NO_2 (>2.5 μl l^{-1} for 8 h). Collapsed and bleached tissues occur mostly at the apex of the leaves and along the margins, most severely on older leaves. With ambient levels of nitrogen oxides, subsequent ultrastructural studies usually reveal reversible swelling of the chloroplast thylakoids similar to that caused by SO_2. Sometimes, NO_2-polluted plants are also greener during early stages of growth but this effect soon disappears.

Crops under glass

Internally-generated nitrogen oxide pollution is a real problem in glasshouses. It is frequent commercial practice in horticulture to burn propane, natural gas, or kerosene to enrich the atmosphere inside the glasshouse with CO_2 as well as to provide heat (see Chapter 8). Optimum benefits of CO_2 enrichment are achieved when levels are raised over three times normal atmospheric levels (about 1000 $\mu l\ l^{-1}$). Since the cost of fuels is an important element in producing a crop, there has been an increasing tendency to vent flue gases from these burners directly into glasshouses. Commercial burners generate NO, sometimes in large amounts ($>1\ \mu l\ l^{-1}$), but little is converted to NO_2 (Reaction 3.19) because O_3 levels are much lower inside than outside the glass. In fact, levels of nitrogen oxides in heated glasshouses with CO_2 enrichment are comparable to those in the most heavily polluted outdoor situation such as to be found in narrow streets with continuous flows of heavy traffic flanked by high buildings.

The levels of NO released by combustion are significant to the health of both plants and grower alike, although their potential to provide for the N needs of a crop is surprising. A flueless heater is capable of releasing 100 kg N ha^{-1} in a 100-day growing season, which is the seasonal glasshouse crop requirement, yet in an unavailable form.

Glasshouse pollution differs in several ways from equivalent outdoor conditions – all of which mitigate damage caused by the high levels of nitrogen oxides. Firstly, there is very little air movement inside a glasshouse and therefore the boundary layers of air above and below leaves are undisturbed. This means that much higher concentrations of nitrogen oxides are required to give similar rates of uptake as are to be found in the turbulent air outside. Secondly, there is proportionally more NO than NO_2 inside glasshouses, which reduces uptake rates because NO is less soluble than NO_2 in extracellular fluids once inside plants. Finally, there is rarely a mixed pollutant situation (see Chapter 11) in glasshouses although unburnt HCs may be a problem. Nevertheless, reductions in crop growth due to presence of high levels of nitrogen oxides are significant in terms of commercial economics. As much as the whole benefit that could have been expected from CO_2 enrichment (see Chapter 8) may be eliminated by the NO.

Prospects

The fact that excessive levels of atmospheric nitrogen oxides can also damage vegetation outdoors is no longer in doubt but how this occurs within the plant is only partially understood. By using normal pathways of N metabolism and inducing levels of either nitrate or nitrite reductase activities (Fig. 3.6), most plants show some evidence of being able to detoxify and utilize the additional N from nitrogen oxides. Nitrate arising from NO_2 is clearly not the problem but the free radical nature of NO may well cause harm. Additional natural capacity to scavenge free radicals again appears to protect plants. In the longer term, these adjustments are often inadequate. The net costs to the plant of general repair and the maintenance of detoxification processes are then reflected in reduced growth.

What can be done to mitigate these effects of nitrogen oxides on vegetation? Clearly legislation to control or limit the amount of emissions from combustion processes is desirable, but this must go hand in hand with technological improvements to burner design, the manipulation of temperature and air/fuel ratios during combustion, and more efficient exhaust trapping of pollutants.

There is another avenue of amelioration which undoubtedly is taking place naturally but could be assisted by plant scientists, especially where a significant loss of productivity or amenity is already taking place in certain species. Vegetation has a vast genetic potential for tolerance against environmental stresses. After light, CO_2 and H_2O, one of the major limitations to plant growth is an adequate N supply. A wide range of inherent adaptations and cellular readjustments take place in most plants to ensure maximum utilization of N under given circumstances. Some are better than others and survive whereas those less able to do so are eliminated. Resistance or tolerance to pollutants is a well-known phenomenon. Indeed, those plants that have enhanced sensitivity to pollutants may be used as diagnostic indicators in the field (see Chapter 1). Until recently, however, tolerance within plants towards nitrogen oxides has rarely been demonstrated. At Lancaster, we now have clear evidence from studies with grasses, cereals and tomato that both natural and artificially mutated plant populations can be screened to produce individuals which are able to flourish at very high levels of nitrogen oxides and take advantage of this additional atmospheric N as a nutrient. This may ultimately prove to be an alternative way of weaning farmers away from excessive use of artificial fertilizers.

Health effects

Industrial accidents

TLVs set for industrial exposures of humans to nitrogen oxides are 25 µl l^{-1} NO or 2 µl l^{-1} NO_2 for a working period of 8 h (Table 1.8). These are well above normal for the public but some unfortunate individuals occasionally experience massive concentrations of nitrous fumes. Unlike sulphuric acid gases, which cause a violent coughing reflex to serve as a warning, nitrous fumes are more dangerous because a fatal quantity may be inhaled without immediate suspicion. Those industries involved in the nitration of various aromatics to form nitrocellulose (the basis of lacquers, films and celluloid) or with nitrophenols used in the drug and dye industries are prone to the hazards of nitrous fumes. The same applies to processes such as metal-etching, photo-engraving, welding in confined spaces or underground blasting. Fires, especially those involving plastics, produce vast quantities of nitrous fumes along with other irritants and hazardous vapours. Even silage or stored hay may cause Silo-Filler's Disease in the agricultural workers nearby. This is due to high rates of anaerobic fermentation or denitrification in the stored fodder which also releases nitrous fumes.

Lack of immediate warning symptoms from nitrogen oxides means that exposure symptoms such as coughing, headaches and chest tightness develop a day later. If particularly severe, they are followed by either sudden circulatory collapse or congestion and water accumulation in the lungs (pulmonary oedema) a short while after this. Treatment with rest, oxygen, steroids and antibiotics is often recommended. However, once initial recovery is made, affected individuals are always likely to have a subsequent history of chronic chest complaints.

Damage is not just confined to the lungs. NO causes the haem of red blood cells (RBCs) to form a type of methaemoglobin. In addition, excess blood nitrate may reduce blood pressure (i.e. act as a vaso-depressant). These, in turn, enhance the destruction of RBCs as well as inducing liver and kidney defects (lesions) as increased efforts are made by the body to eliminate blood breakdown products by means of the urine or the bile duct. Deterioration of the liver and kidneys then often gives rise to jaundice and other related conditions. Unfortunately, in workers (e.g. welders) who do not experience sudden massive doses (acute episodes) but have steady inhalations over a period of time, the chronic illness (Silo-Filler's Disease) they later suffer is often unrecognized and not directly linked with their occupation.

The WHO has recommended a guideline of $0.23 \, \mu l \, l^{-1}$ of NO_2 not to be exceeded for a period of 1 h. Over a period of 24 h, this should be reduced below $0.08 \, \mu l \, l^{-1}$ to safeguard human health.

The domestic scene

Humans in developed countries spend an average of 85% of their time indoors but it is only recently that more attention has been paid to indoor pollution. Indeed, attempts to promote energy conservation have concentrated on restricting indoor–outdoor changes of air which has made the problem worse. Restricted combustion and poor exchange rates also mean that indoor levels of NO predominate over those of NO_2

Table 3.2 Range of NO$_2$ concentrations detected in different rooms of domestic homes[a]

Concentration (nl l^{-1})	Bedrooms (%)	Kitchens (%)	Living rooms (%)
0–21	48	11	40
22–42	33.5	29	49
43–63	13.5	19	18
64–84	2.5	13	5
85–105	2	10	2
106–126	0	7	1
127–147	0.5	3	1
148–168	0	1	1
> 168	0	5	0

[a] Recalculated from Dutch surveys carried out between 1980 and 1982.

– rather like conditions in glasshouses. Even so, levels of NO$_2$ in the home during the winter can be considerable, especially in kitchens and living rooms (see Table 3.2).

For no obvious reason, NO$_2$ has always been assumed to be more toxic to human health than NO, so monitoring studies have concentrated upon NO$_2$ to the complete neglect of NO even though levels of the latter indoors may be higher by a factor of 2 or 3. Such an assumption can only be justified on a short-term basis. By concentrating almost wholly on the effects of NO$_2$ as symptomatic of nitrogen oxides as a whole, animal scientists may well have underestimated the health effects of atmospheric NO. This is compounded by the fact that people in developed countries spend long periods of their lives (85%) indoors in atmospheres containing more NO than NO$_2$.

Levels of indoor NO$_2$ range from 20 to 250 nl l^{-1} with at least as much NO again. On an industrial scale, these higher levels of NO$_2$ are around one-eighth of the industrial TLV but are experienced for far longer than an industrial working day by those who do not venture out quite as frequently – young children, pregnant mothers, and the old – groups most at risk medically anyway. Studies with plants certainly indicate that NO is potentially more hazardous than NO$_2$. This is supported by studies of the effect of NO at a level of 1 µl l^{-1} (often reached as peak values in busy streets) on healthy people, who showed a significant increase in airway and lung constriction after just 2 h.

Added to indoor pollution is the complication of smoking. Cigarette smoke contains nitrogen oxides, especially NO, as well as a multitude of other potential irritants. Some samples of tobacco smoke may contain between 18 and 120 µl l^{-1} NO and similar amounts of NO$_2$. Studies of households with smokers as against non-smokers, or of urban families cooking with gas rather than electricity, have shown higher prevalences of respiratory symptoms in the children of the smoking and gas-cooking households. However, it is only in the latter groups that epidemiological studies have been able to distinguish the medical problems specifically associated with enhanced levels of nitrogen oxides, because there are so many other harmful agents in cigarette smoke which make definitive interpretation of tobacco smoke surveys more difficult.

In many epidemiological surveys, it has also been difficult to establish if nitrogen oxides are the cause of respiratory infections and diminished lung capabilities, or whether they act in concert with other agents, such as SO_2, particulates or photochemical oxidants. Large-scale surveys of lung (pulmonary) function of workers, their children and their parents, some of whom were exposed to daily averages of up to 160 nl l^{-1} NO_2 in the presence of other pollutants, showed no clear links between atmospheric NO_2 levels and lung function. On the other hand, surveys of children and housewives in homes cooking on gas, where hourly peaks of 0.27–1 $\mu l\ l^{-1}$ NO_2 were experienced, showed a positive increase in respiratory illness, while epidemiological studies on other groups, where hourly peaks were much lower (0.08 $\mu l\ l^{-1}$ NO_2), showed no such increased risk.

Healthy and bronchitic volunteers have been exposed to atmospheres containing only NO_2 as well as other pollutants, singly and in combination for up to 2 h. Only levels of NO_2 above 1.6 $\mu l\ l^{-1}$ caused increases in airway resistance or depressions in lung capacity, but no differences between the healthy and bronchitic volunteers have been found. Asthmatic patients, on the other hand, show marked increases in airway resistance to much lower exposures (0.1 $\mu l\ l^{-1}$) of NO_2. Asthmatics often react to a wide range of irritants, including nitrogen oxides, that cause the release of local hormones which, in turn, constrict blood vessels, reduce lung function, and release mucus.

Lung damage

The major difference between air pollution studies on humans and those on plants, microbes and other animals is that there is rarely the possibility of comparable and equivalent biochemical and physiological investigations of humans. Usually, only the pathological consequences may be investigated directly and the use of tissue cultures of human cells or biopsy material is only useful in certain circumstances. Similarly, animal studies must be interpreted with caution when evaluating the hazards presented by similar pollutant gases to humans. Nevertheless, considerable research has been carried out on a wide range of animals to model the human situation and assess the consequences of short-term and long-term exposures of nitrogen oxides. As might be expected, wide differences exist between these models but some salient features do emerge.

Table 3.3　Likely effects of nitrogen dioxide upon humans

Pollutant range^a ($\mu l\ l^{-1}$)	Symptoms
0–0.21	No effects
0.11–0.21	Slight odour detected
0.22–1.1	Some metabolic effects associated with either toxicity, adaptation or repair of lung tissues (e.g. inhibited metabolism of prostaglandin E_2)
1.1–2	Significant changes to respiratory rate and lung volume, enhanced susceptibility to infection and evidence of tissue repair
2.1–5.3	Deterioration of lung tissue (e.g. loss of cilia) not balanced by repair mechanisms
above 5.3	Gross distortion of lung tissues and emphysema, possible death if prolonged

^{a} Over a period of 2 hours (for asthmatics, etc., these values are much lower).

Penetration and retention of NO_2 depends on many factors. In tissue fluids, for example, reactivity of NO_2, its dilution in inspired air, the length of exposure, and the depth and frequency of breathing are all important. As NO_2 is soluble by virtue of Reaction 3.18, it readily enters tissue fluids (mainly as nitrate) much as it enters the extracellular fluid of plants. By the same measure, NO has the potential to enter with rather more difficulty as a mixture of nitrate, nitrite and undissociated nitrous acid. Consequently, the airways passing through the nasal regions, the terminal bronchioles of the lungs, and, finally, the alveoli (Fig. 2.12) all have their surfaces exposed to the products of nitrogen oxides in solution.

During light breathing, about 40% of the nitrogen oxides are taken up in the nose and throat regions but, as exercise is taken and deeper breathing takes place through the mouth, this balance shifts towards the lung tissues which then become the most important uptake surfaces. At low concentrations of NO_2, the most sensitive areas are the junctions between the bronchioles and the alveolar regions. With higher concentrations, the injury extends to the larger airways and into deeper tissues.

Other changes due to nitrogen oxide exposure include accumulations of tissue fluid (oedema), and cellular debris or mucus. There is also a clumping of certain white blood cells (macrophages) which normally accumulate as one of the first lines of defence against infection. Animal lungs have a remarkable ability to rid themselves of inhaled microbes, some of which are pathogenic, and this is essential to keep the alveoli sterile. Low concentrations of NO_2, however, disturb these natural defences of the lungs and cause them to become more susceptible to bacterial attack (see Table 3.3).

Over longer periods, some of the damage caused by NO_2 to the epithelial linings may be repaired if levels of NO_2 are reduced. Newer cells formed during repair, however, are relatively resistant to the effects of the gas by comparison to those originally damaged. Other longer-term studies have also shown that NO_2-induced changes affect blood circulation. The partial pressure of O_2 in blood is reduced in the presence of NO_2 while CO_2 and blood pH levels rise.

At low levels, NO has been found to act as a natural local hormone or modulator produced in the body which reduces blood pressure. In this role, NO appears able to cause relaxation of the smooth muscles surrounding blood vessels. Details of this and subsequent events still remain uncertain but an important regulatory enzyme, guanylate cyclase, is activated by NO. At the moment, the relationship between NO breathed in and the NO produced naturally by the body to regulate blood pressure is unknown.

Table 3.4 is an attempt to summarize most of the essential changes due to NO_2 that have been detected in a range of longer-term animal exposure studies which probably also apply to humans. Unfortunately, the majority of animal studies have been undertaken at atmospheric levels of NO_2 much higher than most people would experience even in the most polluted homes, streets, factories or glasshouses and, consequently, their relevance is questionable. At the moment, it is not possible to produce a comparable table for NO because these studies have not been undertaken.

Metabolic changes

Effects of nitrogen oxides upon lung tissues tend to resemble those of O_3 rather than those of SO_2 or NH_3 because the latter are scrubbed more efficiently from inhaled air by the nasal and throat passages on their way to the lungs. When they finally enter the surface fluid covering the lung alveoli, nitrate or nitrite (as well as undissociated nitrous acid) may coexist (see previous sections) and gain access to cells.

Other N-compounds could also be formed. For example, amino acids and other amines may be converted to more active nitroso derivatives (Reactions 3.35 and 3.37) by undissociated nitrous acid in equilibrium with nitrite. Lung tissue has a poor ability to oxidize nitrite to nitrate or to reduce nitrite to NH_3 which means that nitrosamine ($=N-N=O$) production may be more of a hazard to animals than to plants – especially as most of them are confirmed carcinogens (see later).

$$HNO_2 \Leftrightarrow H^+ + NO_2^- \qquad (3.37)$$

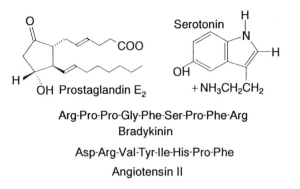

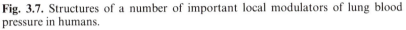

Asp-Arg-Val-Tyr-Ile-His-Pro-Phe

Angiotensin II

Fig. 3.7. Structures of a number of important local modulators of lung blood pressure in humans.

Cellular lipids are susceptible both to autoxidation induced by NO_2 and to *cis* to *trans* isomerizations of their double bonds. *In vivo* simulations in non-aqueous media have demonstrated that at least four different free radical species are generated by attack of NO_2 upon unsaturated linkages in lipids, but the extent to which they occur in living tissues is not known, though the damage they may cause to proteins, containing the amino acid tryptophan, may be important. Deficiencies of vitamin E (α-tocopherol, a natural free radical scavenger synthesized by plants and taken up in the diet) make metabolic disturbances like these more likely.

Autoxidation, promoted by nitroso free radicals, may also form free radicals which react with oxygen to give fatty acid hydroperoxides (see Chapter 6). Natural antioxidants like vitamin E then react with these peroxy radicals rather than with NO_2 to suppress these chain reactions. However, NO_2-catalysed peroxidation differs from O_3-induced peroxidation because the initial peroxyl step is much slower and more readily prevented by antioxidants such as vitamin E.

Both NO_2 and NO form ˙OH radicals (Reactions 3.38 and 3.39) if they react with H_2O_2 which is present in both cigarette smoke and photochemical smog (Chapter 6). ˙OH radicals are so reactive that they attack almost any organic compounds (see Appendix 1) and especially the 1-antiproteinase inhibitor which normally prevents enzymes like elastase from hydrolysing the protein elastin. The elastase then hydrolyses elastin in lung tissues which reduces their elasticity. This means that lung contractions and expansions are restricted exactly like those in the disease emphysema.

$$NO_2 + H_2O_2 \Rightarrow HNO_3 + \text{˙OH} \qquad (3.38)$$

$$NO + H_2O_2 \Rightarrow HNO_2 + \text{˙OH} \qquad (3.39)$$

Table 3.4   Effect on local blood pressure caused by increases[a] in the levels of local modulators

Modulator	Lung blood pressure
Angiotensin II	Increased
Bradykinin	Decreased
Prostaglandin E_2	Increased
Prostaglandin I_2	Decreased
Serotonin (5-hydroxy-tryptamine)	Increased
Nitric oxide	Decreased

[a] Caused by increased rates of synthesis and/or uptake (in addition to reduced rates of breakdown or elimination) of the local modulator. When these are reversed (i.e. local modulator levels fall) then the converse effects on local blood pressure to those listed are obtained.

Changes in blood composition of humans are readily detected and are used to detect and monitor the changes nitrogen oxides have on humans. Both NO and NO_2 react with the iron of haemoglobin to form a type of methaemoglobin which prevents binding of O_2. However, there is no evidence that this type of methaemoglobin arises *in vivo* at the levels of nitrogen oxides to be found even under some of the most adverse working conditions.

Of more relevance are the nitrogen oxide-induced changes in some of the regulatory non-respiratory effectors found in the capillary beds of lungs (Table 3.4 and Fig. 3.7). A number of these modulators or local hormones counterbalance each other and influence local blood pressure within lungs. By so doing they alter the ratio of blood perfused through the tissue relative to the amount of air ventilated through the lungs. Minute changes in blood pressure due to alterations in balance between these modulators have quite significant effects on physiological function of lung tissue because the lung is a low pressure–high flow system. Some modulators or local hormones are made and broken down within lung tissue. Controls on the rates of their synthesis and degradation allow modulation of the levels of some of these local hormones which then exert an even finer control upon their action.

One of these modulators, prostaglandin E_2, has been studied during exposure of rats to low doses of NO_2 (0.2 µl l^{-1} for 3 h). Normally, this prostaglandin causes local constriction of blood vessels which increases blood pressure. It is metabolized to inactive forms which are eliminated from the capillary cells by blood flow. By virtue of this metabolism, the constricting effect of the prostaglandin E_2 is reduced and blood pressure falls. However, when lungs are exposed to NO_2, the metabolism of prostaglandin E_2 is inhibited and, consequently, the vasoconstriction (increased blood pressure) is not removed – an effect which persists for 3 days after removal of the NO_2. The response is therefore highly sensitive to NO_2 and, furthermore, is accumulative.

No doubt there are similar metabolic disturbances of some of the other modulators involved in control of lung function. The fact that NO itself now turns out to be a natural modulator in its own right may be significant and may open the way to a better understanding of respiratory malfunction and the toxicity of the nitrogen oxides.

Dietary implications

The fact that total N deposition rates (wet and dry) to water catchments around conurbations are of a similar order of magnitude to agricultural fertilizer application rates (40 and 100 kg N ha^{-1} a^{-1} respectively) means that atmospherically derived nitrate run-off or seepage into drinking water supplies cannot now be ignored.

Nitrate is not toxic to humans – only when it is converted to nitrite does it cause problems. Blue baby syndrome or methaemoglobinaemia occurs in babies less than one year old because their stomachs are not acid enough to inhibit bacteria which convert nitrate into nitrite. This nitrite then enters the bloodstream and binds tightly to foetal haemoglobin. Fortunately, cases of methaemoglobinaemia are rare but the risks are such that, in areas where the CEC directive for nitrate in drinking water (50 mg l^{-1}) is exceeded, special supplies of low nitrate water should be provided to make up baby milk powders.

The other problem sometimes associated with nitrite is a suggestion of increased incidence of stomach cancers. Theoretically, nitrites react with secondary amines ($=NH$) in the presence of strong acids to form N-nitroso ($=N-N=O$) derivatives. These, in turn, could alter the bases in DNA, causing such affected cells to become cancerous. In fact, epidemiological studies do not give strong support to this theory. Most studies have shown a negative correlation between nitrate exposure and stomach cancer. Even workers in fertilizer plants with high nitrate exposures show no increased incidence of stomach cancer. Very little nitrate is converted to nitrite in the diet and, consequently, any linkage of stomach cancer to enhanced atmospheric deposition of N compounds is highly unlikely.

Further reading

Grosjean, D. *Nitrogenous Air Pollutants: Chemical and Biological Implications.* Ann Arbor Science, 1979, Ann Arbor, Michigan.

Lee, J.A. and Stewart, G.R. Ecological aspects of nitrogen assimilation. *Advances in Botanical Research* **6**, 1978, 1–43.

Marletta, M.A. Nitric oxide: biosynthesis and biological significance. *Trends in Biochemical Sciences* **14**, 1989, 488–492.

Ministry of Agriculture, Fisheries and Food. *Nitrogen and Soil Organic Matter*. Technical Bulletin No. 15, 1969.

National Academy of Sciences. *Nitrogen Oxides*. Committee on Medical and Biological Effects of Environmental Pollutants. 1977, Washington, DC.

Rowland, A., Murray, A.J.S. and Wellburn, A.R. Oxides of nitrogen and their impact upon vegetation. *Reviews of Environmental Health* **5**, 1985, 295–342.

Schneider, T. and Grant, L. *Air Pollution by Nitrogen Oxides*. Elsevier Science, 1982, Amsterdam, Oxford and New York.

Turiel, I. *Indoor Air Quality and Human Health*. Stanford University Press, 1985, Palo Alto, California.

Watkins, L.H. *Environmental Impact of Road and Traffic*. Applied Science Publishers, 1981, London and New Jersey.

Wellburn, A.R. Why are atmospheric oxides of nitrogen usually phytotoxic and not alternative fertilizers? Tansley Review No. 24, *New Phytologist* **115**, 1990, 395–429.

Chapter 4

AMMONIA AND SULPHIDES

'The past is the only dead thing that smells sweet.'
from 'Early One Morning' by Edward Thomas (1878–1917).

Reduced forms of nitrogen and sulphur in the atmosphere

Nitrogen oxides and SO_2 are not the only N- and S-containing gases in the atmosphere. Reduced gases like ammonia (NH_3), hydrogen sulphide (H_2S), and organic sulphides (CH_3SH, CH_3SCH_3, CH_3SSCH_3, etc.) are frequently present and, like the nitrogen oxides and SO_2, form important components of the nitrogen and sulphur cycles. Most atmospheric NH_3, H_2S, etc., is primarily biogenic in origin rather than homogenic, but problems arising from NH_3, H_2S, etc. which are additional to natural atmospheric levels may be traced back to human activities.

Both NH_3 and organic sulphides also have important meteorological implications. NH_3 forms ammonium sulphate particles, when moisture is present, and these form a large proportion of the white hazes which often obscure distant vistas in semi-urban areas. Organic sulphides, by contrast, contribute significantly to general acidification processes (see Chapter 5) and initiate cloud formation, especially over the oceans (see later).

Removal mechanisms of NH_3, H_2S, and the organic sulphides from the atmosphere differ but they have considerable biological importance. This is because these processes influence the amounts of N, S and acidity returned to different ecosystems. Some of these are ill adapted to receive these inputs and respond adversely to them.

Ammonia

Volatilization

Ammonia (NH_3), a highly pungent gas, is a raw material used by industry for the synthesis of ammonium nitrate fertilizer, plastics, explosives, dyes and drugs. It used to have considerable use as a refrigerant. It is still produced in considerable quantities around oil refineries and during the incineration of waste material, especially plastics.

However, local industrial sources of atmospheric NH_3 pale into insignificance on a global scale when biological decay processes are considered. Release of NH_3 from the biological degradation of proteins on soil surfaces (plant residues and animal wastes) into the atmosphere is extensive – a process known as 'NH_3 volatilization'. Atmospheric concentrations of NH_3 in temperate rural regions range from 5 to 10 nl l^{-1} but are much higher near the Equator (around 200 nl l^{-1}). In urban regions, however, levels of NH_3 may rise to 280 nl l^{-1} but much higher concentrations have been recorded (up to 10 μl l^{-1} NH_3) close to industrial and intensive agricultural sources. In some developed countries, atmospheric levels of NH_3 are still rising with accelerated use of artificial fertilizers and higher stocking rates of farm animals.

There are a variety of factors which affect the rate of NH_3 volatilization and some of them are far from simple. NH_3 readily forms cations or complexes of varying stability. Most important among these is the affinity of NH_3 with H_2O to form ammonium ions (NH_4^+) – a process that is strongly enhanced by increased alkalinity (Reaction 4.1). This means that there is an increased likelihood of NH_3 volatilization at high pH.

$$NH_3 + H_2O \Leftrightarrow NH_4^+ + OH^- \qquad (4.1)$$

Levels of CO_2 or changes in temperature also have a marked influence upon this NH_3–NH_4^+ relationship because both the ionization of H_2O and the dissociation of NH_3 are temperature dependent. Furthermore, exchange of NH_3 between solution and the air above varies markedly with temperature.

Reactions of NH_3 other than with H_2O also occur. In soils, NH_3 may be adsorbed onto clay or organic particles and react with carbonyl and other acidic groups to form exchangeable salts. Alternatively, it combines with other organic (particularly phenolic) components to form non-exchangeable products. This means that different soils have different rates of NH_3 volatilization and these, in turn, are affected by their water contents.

Movement of NH_3 as a gas through soils takes place through tortuous diffusion pathways, but in the form of NH_4^+ it may be rapidly transported in solution either by diffusion or by convection if water is moving relative to the soil particles. Generally, the rate of this diffusion increases with both increasing soil water content and NH_4^+ concentrations. Flooding is more complicated because there is a greater chance that the soil beneath is anaerobic which is more likely to carry out denitrification.

Formation of NH_3 from NH_4^+ ions (Reaction 4.1 in reverse) removes

OH^- ions. Therefore, as NH_3 is lost from the soil surface to the atmosphere, the soil solution becomes acidified at rates which depend on the soil buffer capacity. Consequently, volatilization of NH_3 is more likely from soils where the acidity produced can be neutralized by high levels of carbonate or other forms of alkalinity. To a certain extent, this explains why larger emissions of NH_3 occur from natural calcareous soils or after liming.

Rates of NH_3 loss from soils also vary with the anions present in applied fertilizers. Volatilization also increases as the solubility of the non-nitrogenous reaction product decreases. This means that when ammonium fluoride, ammonium sulphate, or di-ammonium phosphate, for example, reacts with calcium carbonate, it forms calcium fluoride, calcium sulphate, or calcium phosphate, respectively – all of which have relatively poor solubilities. Precipitation of these calcium salts then drives Reaction 4.2 in favour of ammonium carbonate which then hydrolyses (Reaction 4.3) to form NH_3, CO_2, and more H_2O. However, if other anions in the applied fertilizer form more soluble calcium salts (e.g. calcium carbonate, nitrate or chloride) then much lower amounts of ammonium carbonate are achieved (Reaction 4.4). This then reduces the amount available for decomposition (Reaction 4.3) and rates of NH_3 volatilization fall.

$$(NH_4)_2SO_4 + CaCO_3 \Rightarrow CaSO_4(\text{ppt}) + (NH_4)_2CO_3 \qquad (4.2)$$

$$(NH_4)_2CO_3 + H_2O \Rightarrow 2NH_3 + CO_2 + 2H_2O \qquad (4.3)$$

$$2NH_4NO_3 + CaCO_3 \Rightarrow Ca(NO_3)_2 + (NH_4)_2CO_3 \qquad (4.4)$$

$$(NH_2)_2C{=}O + 2H_2O + H^+ \Rightarrow HCO_3^- + 2NH_4^+ \qquad (4.5)$$

$$NH_4HCO_3 \Rightarrow NH_3 + CO_2 + H_2O \qquad (4.6)$$

As an added complication, urea is frequently used as an alternative fertilizer because the enzyme urease, which is widely distributed in plants, microbes and soils, catalyses the hydrolysis of urea to bicarbonate and NH_4^+ (Reaction 4.5). As urease activity tends to be greater in soils with large organic contents and rather less in calcareous soils, the usual increase of NH_3 volatilization from alkaline soils and decrease from acid soils is reversed when urea is used as a fertilizer. This enhancement of NH_3 volatilization by urea also occurs extensively with soils treated with animal slurry or wastes rich in urea as the ammonium bicarbonate decays (Reaction 4.6). The observed ratio of NH_3 to CO_2 release ($1:1$) by Reaction 4.6 from slurry-treated fields, however, is half that of Reaction 4.3 on NH_4^+-fertilized land. The cation exchange capacities of these slurry-treated soils are much less in this case because most of the

NH_3 volatilization occurs directly from the slurry surfaces and not from the soil beneath.

Large losses of NH_3 also arise from stored animal manures in stock-yards or from sewage works. The factors involved are similar to those of applied slurries on fields. Periodic addition of fresh material to the tops of piles of manure or the agitation of sewage ponds, therefore, greatly accelerates rates of NH_3 volatilization. Most NH_3 in the atmosphere arises from the direct hydrolysis of the urea in animal urine, other contributions being of less importance.

In Europe, as much as 10% of useful N is lost directly by NH_3 volatilization and, in warmer climates, this can rise to as much as 30%. The amounts of N released into the atmosphere globally by NH_3 volatilization are very large – between 115 and 245 Tg N a^{-1}. Background atmospheric levels of NH_3 over Belgium, Denmark, and the Netherlands, which are intensive arable and livestock-raising countries, are often around 25 nl l^{-1} with peaks up to 75 nl l^{-1} NH_3. Estimates of total N released into the atmosphere from these countries are especially high (300–700 kg N ha^{-1} a^{-1}).

The large releases of NH_3 to the atmosphere over the last three decades are due to (a) increased animal stocking levels, (b) increased human population, (c) increased use of artificial fertilizers, in the form of either NH_4^+ nitrate or urea, and (d) decreased sinks for NH_3 or NH_4^+ uptake. The first three go hand in hand. As material standards improve, humans move from plant-orientated to animal-based diets. This means more food has to be grown to feed animals and this can only be done by using more artificial fertilizer.

Much could be done to reduce N losses associated with applications of N fertilizers due to NH_3 volatilization and run-off of excess nitrate into groundwaters. Direct injection of anhydrous NH_3 or urea at the right depth into soil has not yet been extensively exploited. Even substituting urea for NH_4^+ in irrigation waters (which do not have significant urease activities) will reduce losses due to NH_3 volatilization below 2% in poorer regions where N is unduly expensive, and losses by this route tend to be greater because of higher temperatures.

Removal of ammonia from the atmosphere

Once in the atmosphere, NH_3 neutralizes sulphuric or nitric acids and, by decreasing acidity, promotes the oxidation of SO_2 to sulphate by O_3. Normally, atmospheric NH_3 has an average lifetime of 0.5 h before conversion to NH_4^+. At wind speeds of 10 m s^{-1}, therefore, a molecule of NH_3 travels about 18 km before it turns into NH_4^+.

Measurement of rates of NH_3 deposition are complicated because some intensively farmed lands give off more NH_3 than they receive. On

Table 4.1 Ecosystems thought to be especially sensitive to excessive inputs of nitrogen

Type	Details
Wetlands	1. Ombrotrophic mires (i.e. bog-like areas only supplied with nutrients from rainfall), e.g. raised bogs
	2. Mires in granitic areas or otherwise nitrogen-limited
Lakes	1. Clear water lakes partially covered with certain species (e.g. water lobelia, quillwort and shoreweed)
	2. Nutrient-lacking lakes with a high number of pondweeds including *Potamogeton*
Others	1. Heathlands with high amount of lichen cover with limited mineral availability
	2. Meadows with limited mineral availability used for extensive grazing and haymaking to which artificial fertilizers have never been added
	3. High-altitude coniferous forests

the other hand, fluxes towards damp acidic ecosystems are considerable and cannot be accounted for by stomatal uptake alone as they form perfect sinks for NH_3. For example, the uptake rate of wet heathland in the Netherlands may be as high as 100 kg N ha^{-1} a^{-1}. Ecosystems most sensitive to nitrogen-based atmospheric compounds are those which contain species of plants which are specially adapted to low levels of N and nutrients (Table 4.1). Effects observed in such systems are mainly due to increased competition from faster-growing plant species which were previously restricted in these ecosystems because of the low availability of N. Ultimately, this process leads to the disappearance of the original plant species characteristic of such ecosystems, with an ultimate loss of irreplaceable genetic resource. For example, in bogs only supplied with nutrients from rainfall (ombrotrophic mires), certain mosses (*Sphagnum*) are replaced by seed plants. Similarly, heathlands show a move towards heather-like plants, meadows are threatened by a spread of N-demanding tall perennials like thistles, nettles or willowherb, and lakes grow more algae.

Fears of too much N also reaching natural forest stands by similar processes have also been raised. Indeed, one possible explanation of recent forest decline suggests that too much N from the atmosphere encourages pathogens or causes conifer needles to harden incorrectly (see Chapter 10).

Vegetational changes due to excessive N input also exert changes upon animal populations. Fish are starved of O_2 because of competition with rapidly growing plants. Similarly, vigorous attacks of bark beetles, etc., occur in N-enriched forests.

In most areas where atmospheric levels of NH_3 are high, elevated concentrations of SO_2 also occur. This leads to a phenomenon known as codeposition where fluxes are linked together. One explanation for this is that when NH_3 turns into NH_4^+ (removing acidity) then oxidation of SO_2 into sulphate is enhanced. Removal of SO_2 from the atmosphere as ammonium sulphate at the surface then ensures that the high rates of codeposition of SO_2 and NH_3 are maintained. Similar processes occur in aerosol particles which form clouds. The hygroscopic nature of NH_4^+ and sulphate causes these particles to attract more water vapour so they increase in size and eventually fall as rain, etc. Indeed, wet deposition of NH_4^+ as ammonium sulphate in rainfall is the major removal pathway of NH_3 from the atmosphere.

Effects of ammonia on plants

Most plants are affected visually by atmospheric NH_3 when there has been an industrial spillage and high local concentrations are achieved (> 1.5 µl l^{-1} NH_3). Generally, no visible or invisible effects are detected on even the most sensitive species at levels of 70 nl l^{-1} NH_3 or below. However, there are some claims that deposition of NH_3 and NH_4^+ on certain plants that prefer low levels of N induces leaching of potassium and magnesium which, over the long term, leads to mineral deficiencies.

At higher levels of NH_3 (140 nl l^{-1}), little effect on photosynthesis is detected but long-term exposures (2 months) visibly injure sensitive conifers such as yew, spruces, and cypresses. However, this injury is usually associated with reduced frost hardiness and increased susceptibility to fungal attack. Vegetable and horticultural crops are generally more resistant. Cauliflowers and Brussels sprouts, however, may show characteristic black spots at levels of 0.7 µl l^{-1} NH_3 for 10 days as well as reduced frost hardiness.

Effects of ammonia on health

Atmospheric NH_3 is not normally a hazard to health. Occasionally, industrial spillages release large amounts of NH_3 which immediately react with the moist linings of the throat or the surface of the cornea to form ammonium hydroxide which then causes chemical burns. At much lower levels, the buffering of these layers is normally sufficient to absorb any NH_3 and very little penetrates to the lungs. Irritation of the eyes and throat occurs at levels of 350–700 µl l^{-1} NH_3 – well above ambient atmospheric levels. However, physiological changes have been detected at much lower levels. For example, at 16 µl l^{-1} NH_3, levels of NH_4^+ and urea in the bloodstream are increased. The TLVs vary from country to country. Most are around 25 µl l^{-1} NH_3 for an 8 h day.

Hydrogen sulphide

Malodorous emissions

Hydrogen sulphide (H_2S) – with its notorious 'bad egg' odour – is a highly toxic and flammable gas. There are few industrial uses for H_2S but it is produced in large amounts as a by-product of a number of both natural and industrial processes, especially during the mining of S. Over 90% of global emissions are accounted for by human activities and biogenic emissions of H_2S, by comparison, are quite low. Releases from natural gas or oil reserves, however, are major hazards during the production and refining of high-S fuels, because H_2S is liberated immediately from crude oil or natural gas as soon as it reaches the surface, especially if hot and under pressure.

Decay of organic matter in sewers and breakdown of animal and plant wastes by microbial activity are other major sources of H_2S and organic sulphides (mercaptans). Tanneries, glue and fur-dressing factories, abattoirs, waste treatment plants, and sugar-beet processing all produce significant quantities of H_2S (along with mercaptans) which accounts for the fact that these are some of the least popular neighbouring industries. Tanning is very prone to accidental discharges because the first stage of hair removal uses a paste of sodium sulphide and the following chrome tanning process uses sulphuric acid. If care is not taken to prevent the two effluents from mixing then H_2S is liberated. The manufacture of paper, rayon and S dyes, and the vulcanizing of rubber, are other industrial processes where H_2S may be generated. Special safety precautions are required in all these industries to protect workers.

In areas around pulp mills, levels of H_2S may be as high as 11.5 μl l^{-1} while even higher levels have been recorded inside them (>20 μl l^{-1} H_2S). However, in urban areas, levels of H_2S are normally lower than 5.4 nl l^{-1}. If they exceed this level for any length of time then substantial complaints are likely to be made by the public because our sense of smell with respect to H_2S is very acute.

The human threshold of odour detection for H_2S starts between 0.15 and 1.5 nl l^{-1} but these limits are lower for mercaptans. To a certain extent, sensitivity depends on the individual but the rotten egg smell is clearly apparent to almost everyone at about three times these levels. H_2S is almost always accompanied by other malodorous substances such as methyl mercaptan (CH_3SH), carbon disulphide (CS_2), dimethyl monosulphide (CH_3SCH_3), and dimethyl disulphide (CH_3SSCH_3). The 'quality' of the odour is changed accordingly – presumably from 'bad' to 'worse'.

Microbial exchanges of hydrogen sulphide

Certain S-oxidizing bacteria, characterized by members of the green *Chlorobium* and purple *Chromatium* genera, have the capacity to remove H_2S from the environment. These bacteria of mud and stagnant waters carry out photosynthesis in anaerobic conditions. Instead of splitting H_2O like higher plants and using this process (photolysis) to donate electrons to their photosystems, they split H_2S and release elemental S by a process known as photoautotrophy. Such photoautotrophic bacteria have considerable evolutionary significance, being earlier responsible for the formation of the major proportion of the geological S reserves of the world, but they have little importance to the global cycling of S today.

More unpleasant anaerobic organisms use the amino acids hydrolysed from the proteins of decaying bacteria, plants and animals as a source of C compounds from which they derive energy. As they do so, the S and N of amino acids cysteine and methionine are released as H_2S, CH_3SH and NH_3 without changing their oxidation or reduction state (see Fig. 2.4). These products of a process known as desulphurylation contribute greatly to the stench of putrefaction – a nasty, but necessary, component of the global cycling of S.

Slightly less offensive is 'bad breath' or halitosis. In this case, the ubiquitous anaerobic mouth bacteria multiply when gum disease or tooth decay is prevalent and, as they break down proteins, release unacceptable amounts of CH_3SH. Sweaty socks on unwashed feet also provide an ideal environment for similar organisms to do the same.

Microorganisms of the soil and sewage treatment plants, as well as the plankton of the oceans, produce considerable quantities of CH_3SCH_3, CH_3SSCH_3 and CH_3SH. Estimates suggest that biogenic emissions of CH_3SCH_3 exceed those of H_2S and that a large proportion of these come from the oceans (see later).

Emissions of hydrogen sulphide by plants

Hydrogen sulphide is only slightly phytotoxic. Only at high levels of H_2S (> 100 nl l^{-1}) have necrotic lesions and tip burn been detected on plants. By contrast, beneficial effects on growth have been reported with low concentrations (30 nl l^{-1}) but of more interest is the fact that living, as well as rotting, vegetation is a significant source of H_2S. Indeed, young leaves appear to emit more H_2S than old leaves. When young leaves are exposed to $^{35}SO_2$, more than half is re-emitted as $H_2{}^{35}S$ but the amount of ^{35}S-labelled sulphate remains very low. This means that sulphate is not an intermediate in the synthesis of H_2S from SO_2. The biological implications of the sulphur cycle have already been covered in Chapter 2, but Fig. 4.1 illustrates that production of H_2S from SO_2 is an

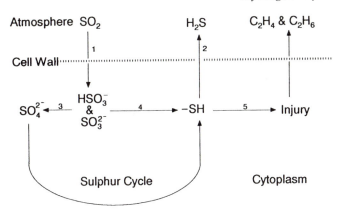

Fig. 4.1. Alternative routes of sulphur metabolism exist for atmospheric SO_2 taken up by plants. SO_2-tolerant plants have either reduced uptake (1) or the ability to metabolize more sulphite or bisulphite to reduced sulphur (4) and to release more sulphur as H_2S (2). This reduces the rate of entry of sulphur from SO_2 into the sulphur cycle (3) which reduces damage in the plant (5) normally indicated by release of ethane formed during lipid peroxidation.

alternative mechanism by which the harmful effects of atmospheric SO_2 upon tolerant plant tissues are dispelled.

The question 'How much sulphur do plants emit as H_2S to the atmosphere?' remains to be answered. Estimates have indicated that of the 100 Tg a^{-1} SO_2 plus H_2S entering the atmosphere by natural processes, more than 7.4 Tg a^{-1} are reliberated by plants as reduced S. Why they should do this is also intriguing. It is clearly not a means of getting rid of excess reductant, because rates of sulphate reduction are less than 0.1% of those of CO_2 fixation, but it may be a mechanism to maintain reservoirs of specific reductants and carriers involved in sulphate reduction in order to maintain a balance between thiol or sulphydryl ($-SH$) and disulphide ($-S-S-$) groups.

Levels of glutathione and cysteine involved in free radical scavenging are carefully controlled in all biological tissues. Perhaps the release of H_2S may be seen rather like a pressure valve whereby excess S is released after S metabolism within cells.

Accidents

Normally, the smell of H_2S is sufficient to warn humans of the presence of this gas at levels below 20 nl l^{-1} and to cause them to retreat or to take remedial action. However, paralysis of the sense of smell occurs at higher levels (150 µl l^{-1} H_2S) so victims may be unaware of the dangers (see Table 4.2). High concentrations of H_2S are actually as toxic as

Table 4.2   Effects of hydrogen sulphide on humans

Exposure range	Effects and symptoms
0–0.2 nl l^{-1}	No discomfort
0.2–1.2 nl l^{-1}	Odour detected
1.2–5.4 nl l^{-1}	Consciousness of 'bad egg' smell
5.4 nl l^{-1}–10 μl l^{-1}	Some discomfort and headache
10–20 μl l^{-1}	Eye irritation threshold
20–50 μl l^{-1}	Severe eye irritation and impairment
50–100 μl l^{-1}	Characteristic eye damage called 'gas eye'
100–320 μl l^{-1}	Loss of sense of smell, nausea and increased lung irritation
320–530 μl l^{-1}	Lung damage and water accumulation
530–1000 μl l^{-1}	Shortage of breath, stimulation of respiratory centre, convulsions, and chance of respiratory arrest
1000 μl l^{-1} and above	Immediate collapse and respiratory failure, followed by death

hydrogen cyanide, HCN. In fact, both have the same TLV of 10 μl l^{-1} and have similar effects. Respiratory failure occurs within seconds due to paralysis of the nervous control of breathing.

The toxicological problems associated with H_2S, like those of HCN, are caused by an inhibition of electron flow through cytochrome oxidase to O_2 in the mitochondria during respiration. The skin of victims of acute intoxication caused by H_2S is usually grey-green in colour as are the internal organs. This is due to the formation of a sulphaemoglobin immediately after death. Such was the fate of 23 inhabitants of the town of Poza Rica, Mexico, in 1950 when a faulty valve on a new gas field installation leaked H_2S for only 20 minutes.

Most of the metabolic consequences of sublethal exposures to H_2S stem from this partial inhibition of cytochrome oxidase in mitochondria. This also increases the number and volumes of RBCs. However, the main clinical effect is irritation of the lung passages and eyes which leaves the victim with pneumonia and conjunctivitis some days afterwards (Table 4.2).

Recovery from acute poisoning is usually without after-effect but some cases show persistent clinical symptoms which result from the initial O_2 deprivation. TLVs for maximum exposure to H_2S set at 10 μl l^{-1} H_2S for 8 h daily exposures would appear to be too high because these levels are also known to cause eye irritation.

Chronic effects due to H_2S are characterized by conjunctivitis, headaches, dizziness, diarrhoea, and loss of weight. Early French medical writers used the term *plomb des fosses* to describe the colic and diarrhoea of Paris sewermen probably from such a cause, which resembled that normally associated with lead poisoning.

Table 4.3 Total global emissions of sulphur [a] from both homogenic and natural sources (adapted from Bates *et al.*, 1992)

Component	Tmol a^{-1}	%
Homogenic [b]	2.40	73.9
Biomass burning [c]	0.07	2.1
Biogenic:		
Marine [d]	0.48	14.8
Terrestrial [e]	0.01	0.3
Volcanic [f]	0.29	8.9

[a] Excludes sea-salt sulphate transfer.
[b] Mainly SO_2 and carbonyl sulphide (OCS).
[c] Not human-induced (the 95% that is initiated by humans is included in the component above).
[d] Oceanic emissions are mainly CH_3SCH_3 (88%), the remainder being H_2S ($< 10\%$), and OCS plus CS_2 ($< 2\%$).
[e] Emissions from land are mainly ($> 60\%$) from tropical regions (20°N–20°S) and consist primarily of OCS (47%), CH_3SCH_3 (27%) and H_2S (20%).
[f] Mainly SO_2 and sulphates ($> 98\%$).

Organic sulphides

Biogenic and homogenic emissions

Several organic sulphides are produced by living organisms (CH_3SCH_3, CH_3SH, CH_3SSCH_3 and CS_2) or as they are subsequently broken down (carbonyl sulphide, OCS). Collectively, these emissions form a significant proportion of the overall global cycling of S even though they amount to only 20% of those emitted by human activities (Table 4.3).

The majority of these organic sulphide emissions originate from marine phytoplankton. Of these, those of dimethyl sulphide (CH_3SCH_3) are by far the most important. CH_3SCH_3 arises from the breakdown of dimethyl-sulphoniopropionate which is produced by the phytoplankton as an osmoticum to regulate cell turgor. The importance of these CH_3SCH_3 emissions from marine sources lies in the fact that cloud formation over oceans is directly linked to CH_3SCH_3 emissions. Cloud condensation nuclei that initiate cloud growth consist mainly of ammonium sulphate aerosol particles which, over the oceans, are derived from CH_3SCH_3. These CH_3SCH_3 emissions from phytoplankton show strong seasonal variations (Table 4.4). In the northern hemisphere, around 50–65°N for example, summer emissions of CH_3SCH_3 rise to 6.5 μmol S m^{-2} day^{-1} but in the winter they are around 1.4 μmol S m^{-2} day^{-1}. Levels of non-sea-salt aerosol sulphate, which form a major portion of the cloud condensation nuclei, mirror this trend. This means that a large portion of global cloud formation, which mainly occurs over the oceans, is linked to the activity of

Table 4.4   Global emissions of CH_3SCH_3 in $\mu mol\ S\ m^{-2}\ day^{-1}$ during different seasons (adapted from Bates *et al.*, 1992)

Region	Summer [a]	Winter [a]
Oceans	54.48	29.68
Land	3.05	2.30

[a] Summer in the northern hemisphere is taken to be from May to October, and winter from November to April. In the southern hemisphere, these periods are reversed.

phytoplankton through their formation of dimethyl-sulphoniopropionate and the emissions of CH_3SCH_3 as it breaks down.

Formation of aerosol sulphate from CH_3SCH_3 involves several reactions involving the ·OH radical in the lower stratosphere. The major products from these reactions are SO_2 and methyl sulphonic acid ($CH_3S(O_2)OH$) – both of which form sulphate after further oxidation. Reactions 4.7–4.9 show one of the atmospheric pathways responsible for methyl sulphonic acid formation from CH_3SCH_3. Methyl sulphonic acid does not arise from human activities and, consequently, is an accurate global indicator of biogenic S emissions. Levels of methyl sulphonic acid over the oceans also vary with the seasons as the phytoplankton blooms appear and fade. Methyl sulphonic acid also contributes significantly to rainwater acidity when marine air masses move across land and deposition subsequently occurs.

$$CH_3SCH_3 + {}^{\cdot}OH \Rightarrow CH_3S(OH)CH_3 \tag{4.7}$$

$$CH_3S(OH)CH_3 + 2{}^{\cdot}OH \Rightarrow CH_3SO{}^{\cdot}H + CH_3O_2 + H_2 \tag{4.8}$$

$$CH_3SO{}^{\cdot}H + O_2 \Rightarrow CH_3S(O_2)OH \tag{4.9}$$

Carbonyl sulphide (OCS) is the major organic sulphide produced by human activities. In many areas, tropospheric levels of OCS often exceed even those of SO_2. In the lower stratosphere, OCS is first oxidized to SO_2 and, later, to aerosol sulphate by ·OH radicals (Reactions 4.10–4.12). Consequently, OCS like CH_3SCH_3 has an effect on cloud formation but this time over land. The major difference between the two is that CH_3SCH_3 is biogenic and has been occurring for some considerable time while OCS is homogenic and, in global terms, has only recently caused change. It has taken some time for the contribution of OCS to spread round the planet. Recently, significant increases in sulphate concentrations between the ice crystals of new snow collected from Antarctica, which can only have come from OCS, have been detected.

$$OCS + {}^{\cdot}OH \Rightarrow CO_2 + HS^{\cdot} \qquad (4.10)$$

$$HS^{\cdot} + {}^{\cdot}OH \Rightarrow SO^{\cdot} + H_2 \qquad (4.11)$$

$$SO^{\cdot} + {}^{\cdot}OH \Rightarrow SO_2 + H^{\cdot} \qquad (4.12)$$

$$CS_2 + {}^{\cdot}OH \Rightarrow OCS + HS^{\cdot} \qquad (4.13)$$

In global terms, carbon disulphide (CS_2) is a very minor S emission. In some instances, however, it is a considerable local pollutant because it is used industrially for the production of viscose rayon fibres. Concentrations of 0.05–0.5 $\mu l\ l^{-1}$ are found near such factories with peak values rising to 1.9 $\mu l^{-1}\ CS_2$. Inside these factories, levels are even higher. Average concentrations remain around 8 $\mu l\ l^{-1}$ even after precautions have been taken. In the open, CS_2 is oxidized in the light to OCS and SO_2 (Reactions 4.13 and 4.10–4.12).

Organic sulphides and vegetation

OCS, the predominant organic S-containing gas in the atmosphere over land, is rapidly absorbed by plants through their stomata. This uptake is light-dependent and a major mechanism by which OCS is removed from the atmosphere. Once inside a leaf, OCS is hydrolysed to H_2S and CO_2. At low concentrations, however, OCS is only slightly phytotoxic and contributes to S nutrition when root supplies are limited.

The dominant organic sulphide emitted by vegetation is CH_3SCH_3. Rates of release from land plants are temperature-dependent like those from phytoplankton. Pronounced seasonal changes are shown at higher latitudes but not during the wet or dry seasons in the tropics. Nevertheless, global emissions of CH_3SCH_3 from land plants are only a small fraction of those from phytoplankton.

Toxicological effects of carbon disulphide

Of all the organic sulphides, only CS_2 presents a significant hazard to humans due to the high concentrations which may exist in and around viscose rayon plants.

Uptake of CS_2 takes place through the lungs into the blood and, from there, into the body tissues. Other possible access routes (skin, food and drinking water) are minor. Once inside, CS_2 distributes itself between lipids (in which it is highly soluble) and proteins to which it binds.

Detoxification involves several mechanisms. It may react with amines to form dithiocarbamates or bind to glutathione (see Chapter 6) before being discharged in the urine. Alternatively, it may react with cytosolic cytochrome P-450 (often invoked by living tissues to detoxify alien chemicals or xenobiotics) to form S and OCS. The latter is then oxidized further to S and CO_2.

Effects of CS_2 upon humans are usually confined to the nervous system, the blood supply, and the eyes. Subacute and acute effects induced by 160–950 µl l^{-1} CS_2 progressively include loss of appetite, irritability, anger, hallucinations, delirium and paranoia. At lower levels (30–160 µl l^{-1} CS_2), changes in colour vision, dark adaptation, accommodation, acuity, and delayed pupil reaction times may be accompanied by conjuctivitis. In the longer term, chronic effects of CS_2 on blood vessels serving tissues such as brain, heart and kidneys have been detected at levels of 6–95 µl l^{-1} along with hormonal disorders.

The threshold for odour perception of CS_2 occurs at about 60 nl l^{-1}. In pure form it appears to be rather sweet and aromatic, but outdoors it is invariably unpleasant because it is accompanied by much lower concentrations of H_2S and OCS to which human sensitivity is very high.

The industrial TLV for CS_2 is very high (20 µl l^{-1} for 8 h) but the WHO has advised a guideline of 32 nl l^{-1} for CS_2 averaged over 24 h below which adverse health effects are not expected. If, however, levels of CS_2 are used as an index of smell around viscose rayon plants then the WHO recommends that this guideline should be reduced to one-tenth of odour perception (i.e. 6.4 nl l^{-1} CS_2 over 30 min).

Further reading

Bates, T.S., Lamb, B.K., Guenther, A., Dignon, J. and Stoiber, R.E. Sulfur emissions to the atmosphere from natural sources. *Journal of Atmospheric Chemistry* **14**, 1992, 315–337.

Freeney, J.R. and Simpson, J.R. (eds) *Gaseous Loss of Nitrogen from Plant–Soil Systems.* Martinus Nijhoff/Dr W. Junke, 1983, The Hague.

Rennenberg, H., Brunold, Ch., DeKok, L.J. and Stulen, I. *Sulfur Nutrition and Sulfur Assimilation in Higher Plants: Fundamental Environmental and Agricultural Aspects.* SPB Academic Publishing, 1989, The Hague.

Rogers, J.E. and Whitman, W.B. (eds) *Microbial Production and Consumption of Greenhouse Gases: Methane, Nitrogen Oxides and Halomethanes.* American Society of Microbiology, 1991, Washington, DC.

Saltzman, E.S. and Cooper, W.J. (eds) *Biogenic Sulfur in the Environment.* Amer. Chem. Soc. Symp. No. 393, 1989, Washington, DC.

Van der Eerden, L.J.M. Toxicity of ammonia to plants. *Agriculture & Environment* **7**, 1982, 223–235.

Chapter 5

ACID RAIN

'When your fountain is choked up and polluted, the stream will not run long, or will not run clear with us, or perhaps with any nation.'
Edmund Burke (1729–1797).

'I do not mean to say that all rain is acid; it is often found with so much ammonia in it as to overcome the acidity; but in general, I think, the acid prevails in the town.'
in 'Air and Rain – The Beginning of a Chemical Climatology' (1872) by Robert Angus Smith.

Formation and deposition of acidity

Definitions

The term '*pluie acide*' was first used by the French chemist Ducros in 1845. But it was Robert Angus Smith who brought the phrase 'acid rain' to prominence in 1872 when he described the acidic nature of the rain falling around Manchester in one of his early reports as the first Chief Alkali Inspector of the UK (see above). The concept that acid rain cannot be considered in isolation from other forms of atmospheric pollution was introduced in Chapter 1, but the intervening chapters have also shown that there are gas-phase reactions which produce acidity in the atmosphere. Such processes are part of 'dry deposition' but gas–water phase reactions followed by precipitation (mist, rain, hail, sleet, or snow) are a part of 'wet deposition'. However, these include both acid- and non-acid-forming processes, so acid rain is only a part of wet deposition, although a very important component. Wet deposition is better described as 'wet precipitation' which, because it is an intermittent event, means that the effects of acidic precipitation are experienced irregularly. Most of the events involved in the wet precipitation of atmospheric pollution are shown in Fig. 5.1.

The formation of droplets in clouds, which then removes atmospheric pollutants as it falls as rain-out, is a very efficient process. Removal rates by rain-out depend on a variety of different factors, including the type and amount of pollutant or the mean size and temperature of raindrops or snow-flakes, etc. Of lesser importance is the process of wash-out beneath clouds,

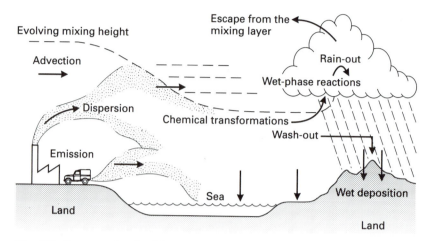

Fig. 5.1. Processes involved in wet and dry deposition (courtesy of The Watt Committee on Energy, UK). By comparing Fig. 5.1 with Fig. 1.1, one may observe how different nations subconsciously perceive their relationship towards cause and effect (clue: look in the middle).

which is more important for the removal of particulates (see Chapter 1).

Emissions of primary pollutants such as SO_2 or NO have already been considered in Chapters 2 and 3. Subsequent gas-phase transformations in plumes or as they escape from the mixing layer (Fig. 5.1) to react with light, oxygen, etc., have been discussed already. However, most of the gas-phase reactions responsible for the formation of acidity outside clouds are relatively slow. Of the four gas-phase reactions described in Chapter 2 (Reactions 2.1–2.8), reaction of SO_2 with $^{\cdot}OH$ (formed from O_3 or monatomic oxygen and H_2O) is the most important (Reactions 2.7 and 2.8). Similar dry gas-phase mechanisms (e.g. Reaction 3.22) exist for the formation of nitric acid from NO_2 but gas-phase oxidation of NO_2 by $^{\cdot}OH$ proceeds 10 times faster than the equivalent reactions involving SO_2.

Gas-phase reactions also generate organic acids. For example, reaction of peroxyl radicals ($HO_2^{\cdot}$) with aldehydes ($HCHO$, CH_3CHO, etc.) leads to the formation of formic, acetic, and higher organic acids which contribute between 5% and 20% towards the total acidity of tropospheric atmospheres.

Cloudwater acidity

The best starting point for consideration of the acid-forming reactions of cloudwater is the question 'What would be the pH[1] of cloudwater if

---

[1] Refer to Appendix 2 for a fuller description of pH, pK_a, etc.

there was no atmospheric pollution?'. If one leaves aside the possible contributions of SO_2 and nitrogen oxides from volcanoes, swamps, lightning strikes, etc., then the major 'natural' gas in the atmosphere contributing acidity is carbon dioxide (CO_2).

$$CO_2 + H_2O \Leftrightarrow H_2CO_3 \qquad (5.1)$$

$$H_2CO_3 \Leftrightarrow HCO_3^- + H^+ \qquad (5.2)$$

$$HCO_3^- \Leftrightarrow CO_3^{2-} + H^+ \qquad (5.3)$$

Carbon dioxide reacts with H_2O to form both bicarbonate and carbonate (Reactions 5.1–5.3). But, as the pK_a of Reaction 5.3 forming carbonate is as high as 10.3 (i.e. when concentrations of carbonate and bicarbonate match), then Reaction 5.2 forming bicarbonate from CO_2 has by far the greatest influence upon the acidity of natural atmospheres. Furthermore, the amount of gas in solution is a function of the partial pressure which in turn is proportional to the mole fraction (i.e. the number of moles of a component divided by the total number of moles in the mixture). In the case of CO_2, this means that if the partial pressure of atmospheric CO_2 is 0.000 35 atmospheres, then Henry's Law constant (K_H) is as follows:

$$K_H = \frac{[H_2CO_3]}{[CO_2 \text{ gas}]} = 3.79 \times 10^{-2} \text{ mol l}^{-1} \text{ atmosphere}^{-1}$$

and the equilibrium constant ($K_{5.2}$) of Reaction 5.2 is given by

$$K_{5.2} = \frac{[H^+][HCO_3^-]}{[H_2CO_3]} = 4.5 \times 10^{-7} \text{ mol l}^{-1}$$

By combining and rearranging these two expressions, one achieves the following equation:

$$[HCO_3^-] = \frac{[CO_2 \text{ gas}] \times K_H \times K_{5.2}}{[H^+]}$$

Thus, if the concentration of bicarbonate in pure water is equal to the H^+ concentration, then by substitution the following is obtained:

$$[H^+]^2 = [CO_2 \text{ gas}] \times K_H \times K_{5.2}$$
$$= 0.000 \ 35 \times 3.79 \times 10^{-2} \times 4.5 \times 10^{-7}$$
$$= 5.97 \times 10^{-12} \text{ mol}^2 \text{ l}^{-2}$$

Therefore
$$[H^+] = 2.44 \times 10^{-6} \text{ mol l}^{-1}$$

and hence

$$pH = -\log[H^+] = 5.61$$

This means that, however, clean the atmosphere and disregarding any wind-blow of alkaline materials, natural cloudwater is always slightly acidic.

Exactly the same approach may be used to calculate the effect of acidic atmospheric gases like SO_2 and NO_2, even though they dissolve and dissociate much more readily than CO_2, because their respective Henry's Law constants and relevant dissociation constants are much greater. Of the various equations for SO_2 (Reactions 5.4–5.6), only the first two are relevant to cloudwater acidity because the pK_a of Reaction 5.6 is around neutrality ($pK_a = 7.2$). Consequently, at pH values less than 5.6, very little sulphite exists.

$$SO_2 + H_2O \Leftrightarrow H_2SO_3 \tag{5.4}$$

$$H_2SO_3 \Leftrightarrow H^+ + HSO_3^- \tag{5.5}$$

$$HSO_3^- \Leftrightarrow H^+ + SO_3^{2-} \tag{5.6}$$

Taking the K_H for Reaction 5.4 as 1.24 mol l^{-1} atmosphere^{-1} and the equilibrium constant for Reaction 5.5 as 1.27×10^{-3} mol l^{-1} then, using very similar calculations to those above, an atmosphere of 100 nl l^{-1} SO_2 produces acidity equivalent to pH 5.9 in clouds nearby due to bisulphite formation. However, 100 nl l^{-1} SO_2 represents a heavily polluted semi-urban atmosphere so other mechanisms beyond the simple dissolution of SO_2 must be involved in the formation of acid rain. This additional acidification by oxidation is described in the following section.

Similar calculations have also been done to assess the natural contribution of the global sulphur cycle (see Chapter 2) to cloudwater acidity which excludes effects due to homogenic emissions or any alkalinity arising from NH_3 volatilization (see Chapter 4), limestone particulates, wind-blow, etc. In these idealized pristine conditions, these calculations still show that cloudwater will be around pH 5.6. Any consideration of oxidation-induced acidification mechanisms, therefore, only needs to take into account those pH-dependent reactions which occur below pH 5.6.

Generation of sulphuric acid

If bisulphite is oxidized to bisulphate within water droplets then the whole relationship between the uptake of SO_2 and the ionization of sulphurous acid (Reactions 5.4 and 5.5) is disturbed and the balance swings in favour of more SO_2 equilibrating with cloudwater and, consequently, more acidity is formed. Moreover, there are other oxidations within cloud droplets which are important acid-forming events in addition to the gas-phase reactions involving $^{\bullet}OH$ radicals that produce acidity (see Chapter 2).

Many oxidants in the atmosphere, such as O_3, hydrogen peroxide (H_2O_2), peroxyacyl nitrates (PANs, e.g. $CH_3COO_2 . NO_2$, see Chapter 5) or other free radicals (e.g. $CH_3O_2^{\bullet}H$ and $CH_3COO_2^{\bullet}H$) do not react with gaseous SO_2 to any great extent. However, when they are dissolved in cloudwater, they readily oxidize dissolved bisulphite ions (e.g. Reactions 5.7 and 5.8) which then forms sulphate and more acidity (Reaction 5.9).

$$O_3 + HSO_3^- \Rightarrow HSO_4^- + O_2 \qquad (5.7)$$

$$H_2O_2 + HSO_3^- \Rightarrow HSO_4^- + H_2O \qquad (5.8)$$

$$HSO_4^- \Leftrightarrow H^+ + SO_4^{2-} \qquad (5.9)$$

$$2HO_2^{\bullet} + M^1 \Rightarrow H_2O_2 + O_2 + M \qquad (5.10)$$

H_2O_2, like O_3 (see Chapter 6), arises from a series of gas-phase reactions (e.g. Reaction 5.10) involving a number of free radicals (see Appendix 1). However, some of these hydrogen peroxide-forming radicals (e.g. peroxyl, $HO_2^{\bullet}$) can also occur in cloudwater droplets and form significant amounts of H_2O_2. The distinction between the two mechanisms (Reactions 5.7 and 5.8) is that oxidation by O_3, although rapid at pH 5, is pH-dependent and falls off rapidly with increasing acidity along with the solubility of SO_2. By contrast, H_2O_2 oxidation is independent of pH so below pH 5 it becomes more important as a means of forming bisulphite in water droplets.

Most of the other free radicals (e.g. PANs) behave like O_3 in this respect although the peroxyacetic acid ($CH_3COO_2^{\bullet}H$) and methyl-hydroperoxide ($CH_3OO_2^{\bullet}H$) radicals resemble more closely the behaviour of H_2O_2. By contrast, O_3 and PAN-dependent oxidations, as well as the pH-dependent Fe^{3+}- or Mn^{2+}-catalysed oxidations of bisulphite by oxygen, fall rapidly as acidity of cloudwater increases. This means that the levels of cloudwater H_2O_2 exert a major effect on acid formation under a wide variety of climatic conditions. Indeed, significant levels of cloudwater H_2O_2 have been detected (0.1 nl l^{-1} H_2O_2 with peaks up to 0.9 nl l^{-1}) by instruments on mountains or carried by aircraft.

[1] See Reactions 2.2 and 2.7.

Table 5.1 Order of importance of different ions contributing to global acid and alkaline depositions

Surface	Acidity	Alkalinity
Land ^{a}	$SO_4^{2-} > NO_3^- \gg Cl^-$	$Ca^{2+} > Mg^{2+} > NH_4^+ > K^+ > Na^+$
Oceans	$Cl^- > SO_4^{2-} \gg NO_3^-$	$Na^+ > Mg^{2+} > Ca^{2+} > K^+ \gg NH_4^+$

^{a} At distances beyond 100 km from the other.

Nitric acid formation

In cold conditions, formation of nitric acid (Reactions 5.11–5.14) takes place in the gas phase (Reactions 5.11 and 5.12) and as N_2O_5 comes in contact with water (Reactions 5.13 and 5.14). By contrast, reaction of NO_2 with ˙OH radicals in the gas phase to form nitric acid (Reaction 5.15) is a warm summer reaction. Different mechanisms operating both in warm summers and in cold winters then explain why nitrate is deposited more evenly over the seasons than sulphate. By contrast, the equivalent gas-phase reaction to Reaction 5.11 for SO_2 (Reaction 5.16) is very slow. This means the warm summer reaction of SO_2 with ˙OH (Reactions 2.7 and 2.8) causes most sulphate to be formed and deposited in late summer and early autumn.

$$O_3 + NO_2 \Rightarrow NO_3 + O_2 \qquad (5.11)$$

$$NO_3 + NO_2 \Rightarrow N_2O_5 \qquad (5.12)$$

$$N_2O_5 + H_2O \Rightarrow 2HNO_3 \qquad (5.13)$$

$$HNO_3 \Leftrightarrow H^+ + NO_3^- \qquad (5.14)$$

$$˙OH + NO_2 + M^1 \Rightarrow HNO_3 + M \qquad (5.15)$$

$$O_3 + SO_2 \Rightarrow SO_3 + O_2 \qquad (5.16)$$

Other sources of acidity and alkalinity

Analysis of ions in rainwater shows clear differences between samples that fall on the land and those on the oceans. Table 5.1 gives the order of importance of different ions which contribute to both global acidity and alkalinity. Sodium and chloride, the dominant ions of seawater, are injected into the atmosphere by an active upward process known as bubble-bursting and carried inland as wind-blown spray. On land, however,

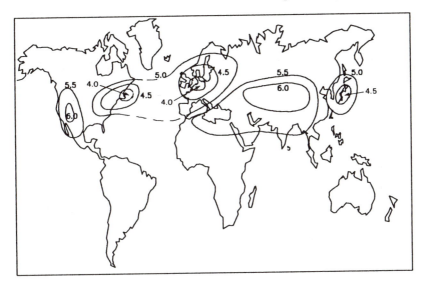

Fig. 5.2. Global ranges of mean rainwater pH (courtesy of Dr D.M. Whelpdale and the WHO Centre on Surface and Ground Water, Canada, and Dr G. Gravenhorst).

natural wind-blow brings alkaline calcium, magnesium or potassium ions into the atmosphere and these outweigh homogenic emissions of fly-ash from combustion or dust from mining and cement operations.

The largest alkaline emissions initiated by human activities are those of NH_3 arising from animal husbandry (see Chapter 4). Loadings of NH_3 in the atmosphere are highly variable and depend on wind speed, the frequency or duration of precipitation, and the nature of the vegetation beneath. NH_4^+ levels in rainwater are highest during droughts or during spring or autumn when ploughing takes place. Elsewhere, the total mass flux of alkaline substances into the atmosphere from cereal-producing prairies is often greater than the mass flux of S caused by emissions from highly industrialized regions (> 50 kg ha^{-1} a^{-1}). This means that some of the rain which falls on the prairies can be distinctly alkaline ($> $pH 8; see Fig. 5.2).

Dispersion and transport

Natural and homogenic emissions to the atmosphere occur by a variety of processes. Apart from volcanoes, most emissions (e.g. from traffic) originate at ground level while other emissions are injected from stacks often over 100 m high. All emissions reach an altitude dependent on the mixing characteristics of the atmosphere prevailing at the time of release. Sometimes at night and all day over snow-covered ground, atmospheres are very stable with little vertical mixing. Under these conditions, stack emissions spread out slowly once they have reached a balanced height with temperature and travel long distances as a coherent plume. However, local topography plays a considerable role. If the stacks are not high enough, emissions are trapped in valleys and flow along watercourses in the manner of liquids.

During the day, heated surfaces and the lower layers of air lift emissions by convection to considerable heights. Plumes may also become unstable and form eddies or loops which assist the dispersion of the plumes. These are often the cause of intense local concentrations of pollution if unstable looping causes the plume to dip to ground level. Again, local topography plays a part. High ground and tall buildings, for example, cause downdraughts which achieve high pollutant concentrations at ground level over quite small areas.

Plumes may be carried away (or advected) by winds and, consequently, climatic factors determine the direction and speed of transport. However, there are also weather conditions that cause the accumulation of pollutants. Slow-moving high-pressure areas with low advection and dispersion rates have the effect of concentrating pollutants. When these eventually meet an advancing low-pressure system, an episode of highly acidic rain may be experienced more than 1000 km or so from the original source. Such high–low pressure transitions are frequent over the eastern USA and northern Europe.

Removal and deposition

Falling rain, snow, sleet, etc., transfer pollutants and their acidic products to land or the oceans by wet deposition. This differs from dry deposition because the rates of dry deposition (D_d) directly depend on pollutant concentration (C_a) and deposition velocity (V_d) which, in turn, may depend on the nature of the land surfaces (see Chapter 1). Rates of wet deposition (D_w), however, do not depend on the underlying surfaces but on the precipitation rate (P), the washout ratio (W, i.e. the concentration of dissolved pollutant per unit mass of cloudwater or rain divided by the concentration of the same pollutant or precursor per unit mass of air, which may range from tens to several thousand), and the ambient air concentration (C_a):

$$D_d = C_a \times V_d \qquad D_w = W \times P \times C_a$$

Figure 5.2 indicates the precipitated acidity over the planet during the late 1970s. Major parts of northern Asia and America show pH values between 5 and 5.5 – very close to pristine conditions (as expected in the absence of homogenic pollution). Areas regularly receiving over 10 times this amount of acidity are those regions along the eastern seaboard of the USA or northern Europe. The greatest annual rainfall acidity (pH < 4) falls on Denmark, southern Sweden and the north of New York State, amounting to 40 times background levels.

Deposition velocities of sulphate ions reach 2.5 g m^{-2} a^{-1} in the most heavily affected regions and those for nitrate amount to an additional third of these values. Indeed, closer to sources and downwind from them, nitrate deposition rates often equal or exceed those of sulphate, especially in northern Europe.

Analysis of rainfall data from the European Atmospheric Chemistry Network over five or more years shows that 29 out of the 120 recording sites had a significant trend of increasing acidity, 23 showed increased deposition of sulphate, and 55 recorded more nitrate. This tendency towards increased acidity was particularly abrupt after 1965.

Instances of rainfall with pH values lower than 3.4 are rare. This is thought to be a lower limit because lifetimes of cloud or rainwater droplets are limited and, at low pH values, solubility of SO_2 is restricted. Moreover, rates of oxidation of SO_2 are much reduced at low pH values because reaction with O_3 is pH-dependent.

Suspensions of acidic droplets in mists at altitude, however, often have pH values much lower than pH 3.4 due to evaporation. Furthermore, these droplets pick up extra oxidants due to their small size (1–100 μm). These highly acidic (pH < 2.75) mists then drift around trees at altitude and adhere to needles by a process known as 'occult deposition'. How much harm they subsequently cause has not been fully established.

Consequences to physical systems

Plant nutrients and soil acidity

Both wet and dry deposition transfer sulphates and nitrates to soils but deposition of protons has the effect of increasing soil acidity. This mobilizes and leaches away nutrient cations which then reduces soil fertility.

Forest vegetation removes pollutants from the atmosphere by direct uptake (see Chapters 2 and 3). There is also an increase in adsorption of ions so that precipitation or throughfall (stemflow when it runs down the trunk) causes an enrichment of certain elements at the base of trees. Sulphate concentrations may be increased by as much as 17 g m^{-2} a^{-1} within certain forest soils by such a process (see Fig. 5.3).

Nitrogen exchanges within a forest are very different. Less than 30% of the N added from the atmosphere appears at the base of trees. Inputs and outputs of S, by contrast, often balance because many of the sparse

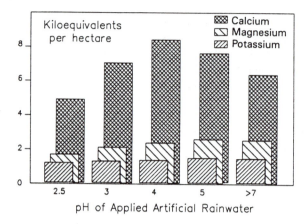

Fig. 5.3. Effect of simulated acid rain of different pH values upon cation content of a Scandinavian forest soil (after Abrahamsen, 1980, cited in Beilke and Elshout, 1983).

Table 5.2   Typical buffer systems operative in most soils

Buffer range [pH]	Buffer system
8–6.2	Calcium carbonate–bicarbonate exchange
6.2–5	Silicates–bicarbonate exchange
5–4.2	Cation exchange systems (e.g. Ca^{2+}, Mg^{2+}, NH_4^+)
4.5–2.8	Hydrated aluminium hydroxide exchange
3.8–2.4	Iron exchange

soils in sensitive areas (e.g. Scandinavia) are fully saturated with sulphate but have very little N.

Under natural conditions, sulphate is often a major anion in soil water. Faster leaching of sulphate increases the loss of cations like potassium, magnesium and calcium. Similarly, increased input of acidity increases the exchange between protons and bound potassium, magnesium or calcium ions. This causes the latter to become more available for leaching. Consequently, those soils most sensitive to acid rain are non-calcareous sandy soils with a natural pH of around 6 and low in colloidal material through which water percolates easily. However, acid soils with low cation content show the largest loss of nutrient ions even though the pH change is minimal. To illustrate this, Fig. 5.3 shows the effect of increasing acidity on the amounts of potassium, magnesium and calcium remaining in soil beneath pine forests.

In most soils, there are various buffering systems which resist changes in acidity. Classic buffering is carried out by a variety of different soil buffer systems, some of which are listed in Table 5.2. The dissolution of calcium carbonate (Reaction 5.17) has been described earlier but carbonic acid–bicarbonate exchange also takes place in various silicates (Reaction 5.18). In more acidic soils (pH 2.8–5), however, there are a series of cation exchange–neutralization reactions which involve the dissolution and precipitation of aluminium. Natural forms of aluminium in rocks and soils are highly variable, existing as silicates, alums, oxides and a wide variety of other forms. Normally, cations like calcium and magnesium are rapidly exchanged from rock and soil by H^+ ions in a reversible and rapid reaction which serves to buffer soil pH (Reaction 5.19).

$$CaCO_3 + H_2O + CO_2 \Leftrightarrow Ca^{2+} + 2HCO_3^- \qquad (5.17)$$

$$CaAl_2Si_2O_8 + 2H_2CO_3 + H_2O \Leftrightarrow Ca^{2+} + 2HCO_3^- + Al_2Si_2O_5(OH)_4 \qquad (5.18)$$

$$[Soil] = Ca + 2H^+ \Leftrightarrow H - [Soil] - H + Ca^{2+} \qquad (5.19)$$

When, however, soils continue to receive large inputs of acidity then this exchange process is limited by the availability of calcium and magnesium in the soil particles. Once these are exhausted then release of Al^{3+} occurs in exchange for H^+ ions. Consequently, acid rain-accelerated weathering releases aluminium hydroxide from the surfaces of soil and rock particles. Aluminium hydroxide has both basic and acidic properties in the presence of strong bases or acids (Reaction 5.20) and is called an ampholyte. Actually, the release of aluminium hydroxide from soil surfaces is more accurately described as a dissolution (Reaction 5.21) followed by a proton transfer (Reaction 5.22). Further transformations (Reactions 5.23 and 5.24) and dissolutions then follow. Consequently, acidity does not disappear when aluminium is mobilized but it is transferred to a progressively weaker cationic acid.

$$Al^{3+} + 3OH^- \Leftrightarrow Al(OH)_3 \Leftrightarrow H^+ + H_2AlO_3^- \quad (5.20)$$

$$Al(OH)_3 + 3H_2O \Leftrightarrow Al(OH)_3(H_2O)_3^0 \qquad (5.21)$$

$$Al(OH)_3(H_2O)_3^0 + H^+ \Leftrightarrow Al(OH)_2(H_2O)_4^+ \qquad (5.22)$$

$$Al(OH)_2(H_2O)_4^+ + H^+ \Leftrightarrow Al(OH)(H_2O)_5^{2+} \qquad (5.23)$$

$$Al(OH)(H_2O)_5^{2+} + H^+ \Leftrightarrow Al(H_2O)_6^{3+} \qquad (5.24)$$

If excess acidity still remains then mobilization of iron oxides may also occur as hydrated forms of Al^{3+} ($Al(OH)^{2+}$ and $Al(OH)_2^+$) are lost to run-off waters draining into streams, rivers and lakes. Such processes

Table 5.3 Annual heavy metal exchange by a spruce forest ecosystem at Solling, Germany (after Meyer – see Beilke and Elshout, 1983)

Element output	Deposition $[kg\ ha^{-1}\ a^{-1}]$	Soil seepage $[kg\ ha^{-1}\ a^{-1})$
Aluminium	2.8	24
Cadmium	0.02	0.03
Copper	0.66	0.11
Chromium	0.17	0.006
Iron	2.1	0.16
Lead	0.73	0.013
Nickel	0.14	0.07
Zinc	1.7	2.4

occur very quickly in hard granitic regions (as in southern Norway) which experience large amounts of acidic rain. Coupled to this are the very low weathering rates of granite which provide little exchangeable calcium or magnesium to counteract this acidity. Plate 8 shows similar sensitive areas in the UK expressed in terms of the critical loads that may be tolerated (see Chapter 1).

Soilwaters, therefore, involve complex processes which interact with acidity. Essentially, these are of 3 types – exchange, buffering and neutralization. During exchange, protons replace adsorbed basic cations (calcium, magnesium, potassium, sodium or ammonium) or take up new exchange sites. If there is a strong counterbalancing anion like sulphate then basic cations are lost in run-off. Soil buffering (see earlier) involves, for example, exchange reactions between bicarbonate and aluminium ions (Reactions 5.17 to 5.24).

Formation of salts from acidic inputs reacting with bases in the soil contribute to neutralization. Many factors are involved in neutralization but soil composition and temperature are the most important. Eventually, after sustained acidic inputs, available bases within certain soils are reduced but, in certain circumstances, acidification increases the rate of mineral weathering. Consequently, either fresh release of calcium and magnesium takes place, or the products of weathering and neutralization build up close to exchange sites which prevents a further increase in weathering rates and, with time, may even decrease weathering rates. Clearly, with so many soil types and conditions, it is difficult to extrapolate these micro-effects within soils to the wider global context.

Acidic groundwater is also known to cause considerable dissolution of buried metal pipework. Furthermore, some heavy metals which are not usually very mobile are encouraged to dissolve by increased acidification. Normally, they bind strongly to organic material and only show mobility when humus levels are low and levels of acidity especially high. Table 5.3 shows how different heavy metals change in a spruce forest soil with time. All heavy metals tend to remain in soils with the exception of aluminium and zinc. Lead, which is widely distributed in the environment (see Chapter 9), also shows little inclination to emerge in run-off waters.

Finally, mention must be made of the NH_4^+ ions which reach the soil in raindrops or aerosols containing ammonium sulphate, nitrate or carbonate. This input of NH_4^+ ions is directly opposed to that of NH_3 volatilization (see Chapter 4), and the equation governing the equilibrium between NH_3 and NH_4^+ (Reaction 5.25) is heavily in favour of NH_4^+ formation and against release of NH_3 at soil pH values lower than 8. This is despite the fact that the pK_a of Reaction 5.25 varies with temperature, humidity, and CO_2 concentration. According to Reaction 5.25, NH_3 volatilization causes soil acidification but, fortunately, this acidification is only a problem of alkaline soils with a high calcium content (i.e. lime-treated). The major effect of additional inputs of NH_4^+ from rainfall and aerosols to acidic soils is to provide an extra basic exchangeable cation to assist calcium and magnesium with neutralization.

$$NH_3 + H^+ \Leftrightarrow NH_4^+ \qquad (5.25)$$

There is evidence that changing agricultural and forestry practices in remote regions also contribute to groundwater acidification. In the past, upland pastures were often heavily manured in spring and the NH_4^+ added helped to neutralize groundwaters. With rural depopulation, many of these pastures have been abandoned and the practice is much reduced. Intensive forest plantings have also increased the amount of needle litter. These are potent sources of acidity in their own right and increase groundwater acidities still further.

Evaporation and sublimation

There are a number of mechanisms by which precipitated acidity causes damage or deterioration. If rainwater falls on vegetation or rocks, it may either flow away immediately, enter a fissure, or remain held as a droplet by surface tension (Fig. 5.4), especially if the surface is hydrophobic (water-repelling) like the cuticle of a leaf. Leaves or conifer needles also have pits (e.g. sunken stomata) or hairs which catch and retain droplets, especially if mist droplet sizes are small.

Such droplets or fluid-filled crevices are subjected to a variety of microclimates. For example, the water in them may freeze or evaporate which affect their ionic composition. As water molecules are lost, the concentration of ions (including protons) rises, causing local pH to fall, sometimes to quite low levels. This means that high localized concentrations of acidity can then attack surrounding structures. These microloca-

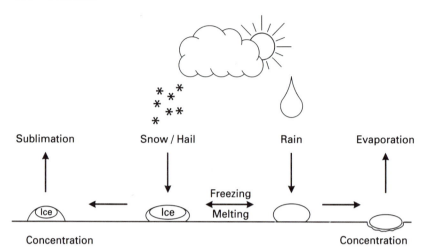

Sublimation   Snow / Hail   Rain   Evaporation

Freezing

Melting

Concentration   Concentration

Fig. 5.4. The various processes involved in the concentration of acidity upon leaf and rock surfaces.

tions, in turn, provide focal points for repetitions of these sequences if precipitation reoccurs. This has the effect of focusing damage upon critical points (Fig. 5.4).

At least two other acid-concentrating processes also occur within droplets. During ice formation, water molecules sweep other molecules in the droplet ahead of the forming ice crystals. This means that elevated levels of other ions, etc., are to be found only at the ice crystal boundaries. Consequently, if the temperature fluctuates around 0°C and this process is repeated many times, this sweeping effect concentrates acidity in certain microlocations. However, if the air above the frozen droplet or fallen snow particles is warmer or drier, sublimation also takes place (Fig. 5.4). This occurs when water molecules in the solid phase move directly into the gaseous phase leaving behind a greater concentration of ions (including H^+). However, the immediate consequences of ice crystal sweeping and sublimation are less than those of unfrozen droplets because lower temperatures reduce rates of attack on the surface beneath. Once melted, however, the effects are sudden and more damaging because highly localized levels of acidity are achieved.

Water droplets do not even need to land on a surface to undergo such acidification. Aerosols or mist droplets may be acidified at high altitude by similar combinations of enhanced oxidation, evaporation, and sublimation. As a result, when such mists come in contact with vegetation during occult deposition (droplet interception), they are already more acidic than rainwater. These have their greatest effect at the tips of conifer needles or on the upper branches or crowns of trees. Some very

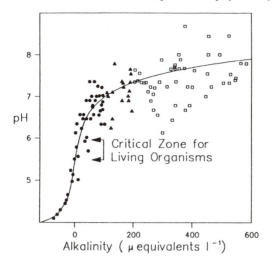

Fig. 5.5. Relationship between acidity and various types of hardness to be found in the upland lakes and tarns of the Lake District, UK, in 1983. Symbols: permanently acid lakes, circles; soft waters, triangles; medium hard waters, squares (courtesy of Drs Sutcliffe and Carrick and the Institute of Freshwater Ecology, Windermere).

low determinations of pH (< 1) have been recorded in ice/water droplets on conifer needles in central Europe where repeated evaporation–sublimation cycles are known to have taken place either on the needles or in the mists that swirl around them.

Acidity in rivers and lakes

Bicarbonate is the most important component in the neutralization of freshwater. High concentrations of bicarbonate occur in 'hard' waters, but in waters below pH 5.4 bicarbonate is completely absent. The description 'permanently acid' is used to discriminate acidified waters from soft waters which do contain small amounts of bicarbonate. There is also a biological distinction between permanently acid and soft waters because the pH zone between 5.7 and 5.2 marks a distinctive boundary between survival and elimination of many freshwater animals (Fig. 5.5).

Many studies have demonstrated the increasing proportion of permanently acid lakes in sensitive areas in Scandinavia or North America. Table 5.4 illustrates typical trends towards acidification in southern Norway. In practice, it is difficult to calculate rates of acidification of lakes because climatic, geographical, and geological factors are so variable. However, over the last 30 years, analyses of Arctic ice have

Table 5.4   Changes in levels of pH in 87 lakes in southern Norway over time (after Wright, SNSF Report TN 34/77, 1977, p. 71)

| | *Number of lakes [expressed as a percentage]* | |
pH range	*1923–1949*	*1970–1976*
4.5 and below	0	2
4.6–5	3	25
5.1–5.5	18	14
5.6–6	21	17
6.1–6.5	11	15
6.6–7	20	13
7.1–7.5	17	9
7.6–8	3	5
Above 8	7	0

shown that there has been a regular decrease of 0.007 pH unit a^{-1} which is directly attributable to homogenic emissions. Similar trends are now emerging from measurements on Antarctic snow. Measurements of Arctic polar ice also show that greatest acidification occurs during spring. This coincides with increased movement of polluted air into the polar regions at the end of the polar winter. In summer, these regions are protected by the polar weather fronts which occur well to the north of industrialized regions.

Measurement of pH alone is not an accurate description of the acidic properties of natural waters. The H^+ ion concentration of acidified lakes depends not just on the concentration of carbonic acid but also on the concentrations of strong bases, of other weak organic acids and, above all, on the concentration of the deposited strong acids. These are technically difficult and require correction for volume changes during titration. Influences of the atmosphere above have also to be taken into account. If one regards the H^+ ion concentration as the non-neutralized residue from sulphuric and nitric acids then gaseous concentrations of SO_2 and NO_2 as well as relative humidity must be known.

The major neutralizing agent which enters lakewater from the atmosphere is NH_4^+. Consequently, knowledge of the gaseous concentration of NH_3, as well as all the aqueous components, is relevant if it is important to establish how far a particular atmosphere–lakewater system is from equilibrium. Not surprisingly, large equations and extensive calculations are required. From studies so far, it has been shown that at least two organic acids, as yet not positively identified, with pK_a values of 5.65 and 3.55 contribute significantly to the acidity of many lakewater systems.

Plant effects

Leaf injury and cuticular weathering

Injury to foliage as a direct consequence of acidic wet deposition remains controversial. Visible damage to leaves of sensitive crop plants like radishes, beets, soya beans and kidney beans has been observed when exposed to artificial rain of pH 3.4 or lower – unlikely though this is in natural conditions. The degree of injury by acid rain depends on dosage – a function of concentration and period of contact of acidic droplets on leaf surfaces. Factors such as temperature, humidity, wind turbulence, surface wettability, and leaf morphology, all influence the period of contact and, therefore, the extent of injury. On sensitive plants, 95% of all foliar lesions due to acidic precipitation occur where water normally accumulates on leaves – along leaf veins or margins, near hairs and close to stomata. Strong acids may then hydrolyse the waxy esters of cuticles to release long fatty acids. This then changes the water-repelling or hydrophobic characteristics of leaf cuticles and increases their wettability.

Field-grown plants appear to be much less susceptible to the development of foliar symptoms than those treated with simulated acid rain. Abraded waxy cuticles and other changes in conifer needle structure, however, are often detected even though no specific changes due to particular precipitation events can be identified. The acid-induced cracking of the thin waxy plugs which cover the stomata of conifers also allows penetration of pollutants, enhanced pathogen attack, increased water loss, or more frost damage. However, it would appear that direct effects of acidic wet deposition on leaves and needles alone are not an explanation of forest decline (see Chapter 10), but it may be an additional stress which adds to the long-term reductions in the growth of trees and crops.

Physiological and biochemical changes

Studies of changes due to acidic wet deposition, as distinct from those caused directly by dry deposition of SO_2 (Chapter 2) or nitrogen oxides (Chapter 3), are rare. There are reports of reduced pigment levels, but no changes in rates of photosynthesis or respiration have been detected unless unrealistic levels of pH (<2) have been used. This makes it very difficult to attribute reductions in growth of sensitive crops like radishes, beets, maize and soya beans directly to acidic wet deposition over the pH range of 4–5.

The importance of adequate *in vivo* buffering or 'pH-stat' mechanisms has been emphasized in Chapter 2. Certain plants resist cellular pH change, even when exposed regularly to simulated rain down to pH 3, by pumping the extra acidity into their vacuoles. The size and nature of buffering capacity is an inherited characteristic. Consequently, acid-tolerant plants resist pH change while acid-sensitive plants are less adaptable. It is possible that the energy redirection hypothesis (see Chapter 1) is one way of explaining net growth reductions of crop plants due to acidic wet deposition. As more pressure is put upon cellular systems to resist pH change, more energy is devoted to make the 'pH-stat' proton pumps work harder to push H^+ ions outwards from cells (or across the tonoplast membrane into the vacuoles). This means there is less energy for growth.

One feature of simulated treatments is that it is the composition of the experimental rain and not the overall acidity that determines response. Mixtures of sulphuric and nitric acid have more effect than, for example, nitric acid alone. Moreover, the response of the plant to mixtures of sulphate and nitrate is similar to that associated with free radical-based attack. Moreover, levels of free radical scavengers increase and cellular damage is signalled by release of hydrocarbons like ethene and ethane. This indicates that part of the failure to grow is due to mechanisms of injury which are similar to those normally associated with O_3 (see Chapter 6).

Foliar leaching and reproduction

Application of simulated acidic rain on plants releases cations like potassium, magnesium and calcium from leaves which then accumulate at the bases of plants along with other nutrients. Enhanced uptake of such nutrients by roots often offsets such losses. However, seed germination, seedling growth, flowering, and seed formation are sensitive to simulated acid rain treatments, but little is known about long-term effects on flowering or fruiting. The full significance of this for forest regeneration has yet to be determined but, undoubtedly, any individuals that show sensitivity to acidity at early stages of development are preferentially eliminated.

Animal effects

Lakes and fisheries

Following acidic precipitation, evaporation and sublimation processes concentrate run-off as it flows over vegetation and down tree trunks, and percolates through soils. Decaying vegetation also contributes additional acidity as microbial activity breaks down organic debris both on the surface and deeper in the soil. Subsequent exchange of ions between

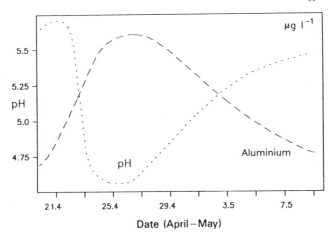

Fig. 5.6. Changes in levels of acidity and aluminium content of a Scandinavian river system during a spring flush event.

soil particles is extensive and may alter particle properties so that the water table levels change. If waterlogging follows then microbial anaerobic reduction is promoted and more acid is formed (see Chapter 2). Drying out of soils, however, reoxidizes sulphide to sulphate. Similarly, conversion of nitrate to NH_3 reduces acidity.

In most acid-affected areas, there are only small differences ($< 10\%$) between inputs of S compounds into the catchment area and the amount of S appearing in run-off (i.e. the whole system becomes saturated). Barren bedrock with little soil, typical of many areas in Scandinavia, shows little capacity to retain S compounds. However, peaty soils show input and output differences of S of more than 40%.

There are also wide variations in acidic run-off. As streams flow through limestone or chalk they become less acid, while lakes show pH variations both with depth and between inlet and outlet streams. Seasonal variations are pronounced, especially in rivers. Snow-melt is often most closely associated with high levels of acidity. Figure 5.6 shows a typical experience in Scandinavia river systems. A fall in pH immediately after snow-melt is followed by an increase in the levels of aluminium ions released from the soil/rock systems as a consequence of this acidity.

Short-term effects of acidity upon fish are listed in Table 5.5. However, there is considerable variation in sensitivity between species (and even individuals) at different life cycle stages. Other factors besides acidity are also important. Levels of bicarbonate, phosphate, iron, and trace metals all cause variations in response.

The most critical ions affecting fish health, other than H^+ and phosphate ions, are undoubtedly those of aluminium. Many studies have

Table 5.5 Short-term effects of acidity upon fish (after Alabaster and Lloyd, *Water Quality Criteria for Freshwater Fish*, pp. 21–45, FAO, Butterworths, London, 1980)

pH range	Effect
6.5–9	No effect
6.0–6.4	Unlikely to be harmful except when CO_2 levels are very high (>1000 mg l^{-1})
5.0–5.9	Not especially harmful except when CO_2 levels are high (>20 mg l^{-1}) or ferric ions are present
4.5–4.9	Harmful to the eggs of salmon and trout species (salmonids) and to adults when levels of Ca^{2+}, Na^+ and Cl^- are low
4.0–4.4	Harmful to adult fish of many types which have not been progressively acclimated to low pH
3.5–3.9	Lethal to salmonids, although acclimated roach can survive for longer
3.0–3.4	Most fish are killed within hours at these levels

shown that aluminium ions are toxic to fish and this toxicity is related to the level of acidity. Studies on trout, for example, have shown that levels of 100 µg l^{-1} aluminium (the highest level recorded during spring flush events like that shown in Fig. 5.6) allow trout to live for more than 14 days when the pH is maintained at 5.2. However, an increase of acidity (to pH 4) with levels of aluminium unchanged causes most fish to die after five days even though no gill damage is observed.

Calcium levels, in the presence of aluminium ions, are critical If calcium concentrations are low then aluminium becomes toxic, but when calcium levels are high then, even at low pH, aluminium ions have no effect. The problem is quite complex because newly hatched fry are most sensitive to acidity whereas the later 'swim-up' fry are more sensitive to aluminium ions.

There is no doubt that the long-term acidic precipitation causes freshwater fisheries in affected areas to decline. In 1983, for example, it was estimated that lakes in southern Norway representing over 13 000 km^2 of surface area were devoid of fish. Figure 5.7, showing the proportion of lakes which have lost their brown trout over a period of 35 years, is taken from a survey of 2850 lakes in southern Norway, and illustrates the extent of the problem.

Another long-term problem also occurs even if liming of affected waters is undertaken at regular intervals. Phosphate is a critical nutrient for fish growth and levels of phosphate in acid-affected waters remain persistently low. It appears that aluminium ions precipitate available phosphate as insoluble aluminium phosphate in soils very close to streams before they run into lakes. Remedial liming of lakes to limit acidification should therefore also include extra phosphate.

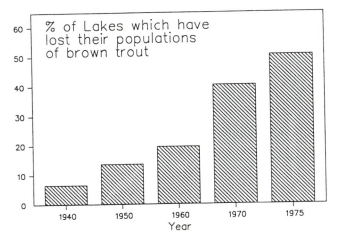

Fig. 5.7. Proportions of Scandinavian lakes (2850 in total) which have lost their populations of brown trout.

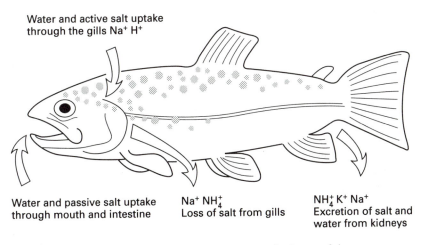

Water and active salt uptake
through the gills Na⁺ H⁺

Water and passive salt uptake   Na⁺ NH₄⁺   NH₄⁺ K⁺ Na⁺
through mouth and intestine   Loss of salt from gills   Excretion of salt and
                                                                water from kidneys

Fig. 5.8. The various exchanges of ions important to freshwater fish.

Mechanisms of toxicity in fish

The blood of freshwater fish contains high levels of both sodium and chloride (equivalent to 150 mm l⁻¹ NaCl) and lesser amounts of calcium, potassium and bicarbonate. Soft fresh waters, however, contain less than 0.1 mm NaCl. The fish that swim in them must therefore replace any losses of NaCl in the urine or from the gills by uptake through the gills (Fig. 5.8).

Fish gills have a complex structure. Gill slits divide the sides of the throat into distinct gill arches. Gill filaments project from these arches

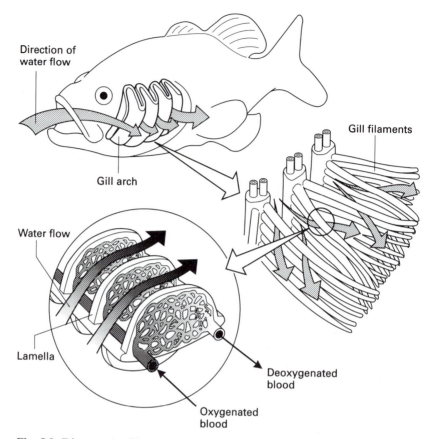

Fig. 5.9. Diagrams to illustrate the structure of fish gills and the various blood supplies that serve them (adapted from Campbell *Biology* (Second edition), Benjamin/Cummings, 1990).

and each of these are separated into individual gill lamellae (Fig. 5.9). The majority of the cells covering the gill lamellae either produce mucus or exchange O_2 in respiratory cells. There are also separate areas in gills covered with 'mitochondria-rich cells'. Respiratory cells are supplied by arterial blood and take up sodium and chloride. Mitochondria-rich cells, by contrast, are served by the venous system and recover calcium.

Uptake of chloride and sodium are separate processes and both require energy in the form of ATP. Meanwhile, OH^- and bicarbonate ions are counterexchanged outwards as chloride is drawn in. Sodium pumps of fish in natural waters, however, are able to capture sodium against a 1000-fold difference when the levels of H^+ ions are the same both inside and out. In acid waters (pH *ca.* 4), the counterbalancing H^+ ions must be expelled against a 1000-fold gradient and the workload is therefore greater. Because large concentration gradients exist in both directions, advantage is always taken of NH_4^+ ions inside and these are ejected instead of H^+ ions. Gill membranes are partially permeable to sodium and highly permeable to H^+ ions. Even if the latter are ejected by active pumping they return passively (i.e. without energy expenditure). Calcium ions, however, have the ability to reduce this easy access. In other words, at high calcium levels, less H^+ ions enter and less sodium ions leave.

Meanwhile, bicarbonate, one of the counterions for chloride uptake, is converted to CO_2 if acidity rises but this also makes the active entry of chloride more difficult. Calcium, however, also reduces the chloride permeability of gill membranes. Consequently, chloride is lost and more protection against acidity is afforded.

The primary cause of fish death in acid waters (Table 5.5) is excessive loss of critical ions like sodium which are not replenished by pumping. When blood plasma sodium and chloride levels fall by about a third (to the equivalent of 100 mM NaCl), body cells swell and fluids surrounding them are concentrated. Loss of potassium from cells may partly compensate for these changes, but if the potassium is not eliminated from the body fast enough then depolarization of nerve and muscle cells takes place. Uncontrolled twitching of non-acclimated fish suddenly exposed to acid waters is symptomatic of these effects.

It is therefore calcium which enables fish to cope with acid waters and allows them to acclimatize. Calcium ions naturally bind to external surfaces of gills. Once fish are adapted to acidic waters, the affinity of calcium for the gill surfaces increases. However, sulphate is more effective than nitrate or chloride in displacing calcium from binding sites at pH values above 3.8. Below this there is very little difference.

Calcium is one of the major regulator molecules in the cells of all living systems even at very low levels. The modulation of membrane permeability as described is just one of many cellular events controlled by free calcium ions. However, extra aluminium ions interfere with this calcium-regulated permeability. As stated earlier, aluminium is toxic to fish at certain pH levels around 5 to 5.5 but less so at higher or lower levels of acidity. Spring flush events during snow-melt (Fig. 5.6) also have pH values around 5 and high concentrations of aluminium ions in the form of monovalent $Al(OH)_2^+$ (more accurately $Al(OH)_2(H_2O)_4^+$). This promotes the efflux of sodium ions at these pH levels, but if the pH is lower then more $Al(OH)^{2+}$ and Al^{3+} ions are present and the losses of sodium are less. The toxicity of $Al(OH)_2^+$ ions is related to the fact that they cause mucous clogging of gills, interfere with respiration, and inhibit other regulatory events associated with calcium. The ionic form of aluminium also inhibits developmental events such as the calcification of the skeleton as fish fry mature. Losses of immature fish fry due to these causes and their failure to grow through to maturity are responsible for the long-term loss of fish stocks.

Effects on invertebrates and birds

Mayflies, caddisflies, freshwater shrimps, limpets, snails and beetle larvae are also absent from acidic waters when the pH falls below 5.5. Decreases in their numbers due to acidity do not account for the decline of fisheries. Invertebrate populations have declined in parallel with those of fisheries not ahead of them. The mechanisms are similar: disturbances in pH control and loss of osmotic regulation due to calcium and aluminium imbalances – exactly the same as for fish.

Aluminium and heavy metals released by acidification of run-off waters are concentrated by many invertebrates and enter several food chains. Those birds which are reliant upon freshwater invertebrates receive secondary doses of aluminium and heavy metals in their diets. Bird numbers are then reduced in such areas due to imperfect calcification of their eggs which fail to produce live fledgelings.

Indirect effects upon humans

There are no known direct effects of acid rain on humans. Occasionally, however, people experience sudden industrial exposures to aerosols or mists containing commonly used acids such as sulphuric, nitric, hydrochloric and hydrofluoric acids.

Accidental exposure to acidic aerosols is a well-recognized hazard. The symptoms are, by and large, similar to those of SO_2 (Chapter 2), NO_2 (Chapter 3), and HF (Chapter 9). All induce irritations and pulmonary disturbances. The main differences arise from the size of the aerosol droplets. Droplet sizes of 0.8 μm or less are the most harmful. In other words, more lung damage is caused by a lesser amount of acidity if it is more finely divided.

Indirect effects of acid rain have been suggested to be a hazard to certain human populations over long periods through changes in water quality. By far the largest contamination of natural waters involves nitrate. The majority of this arises from excess run-off from fields treated with artificial fertilizers which reappears in aquifers used for drinking water supplies. Acidic precipitation upon such water catchments has little, if any, impact upon the extra inputs of nitrate.

The link between high-nitrate water supplies and the occurrence of stomach cancer is tenous. Certain mouth bacteria turn nitrates into carcinogenic nitroso $(=N-N=O)$ derivatives and individuals with unusually high nitrate diets do tend to have higher rates of stomach cancer. The CEC has limited nitrate levels in public water supplies to a maximum of 50 mg l^{-1}. Unfortunately, levels in drinking water in the Netherlands and certain areas of the UK often exceed this level. The most favoured mechanism for nitrate removal is dilution with low nitrate water from elsewhere, but it is also possible to use ion exchange to replace nitrate with bicarbonate followed by microbial denitrification to reduce the exchanged nitrate to nitrogen.

The most likely hazards from alterations to public water supplies by acidic precipitation are increases in the aluminium leached from soils in certain water catchment areas. The number of cases of osteomalacia, a rare bone-wasting disease, has risen in areas where aluminium levels in water supplies are high ($1000–2000 \, \mu l \, l^{-1}$). A possible link between the two was drawn when patients suffering from kidney failure were once treated with aluminium which had the side-effect of bone softening. This, in turn, could be alleviated by the use of an aluminium-chelating drug, desferrioxamine, which is also effective in treating osteomalacia. Such interference of aluminium with calcification of bones in humans is analogous to the imperfect calcification of fish fry when they fail to mature in acidified, aluminium-enriched waters.

Another disease attributed to the availability of aluminium in natural waters is Alzheimer's disease, a senile dementia that occurs much earlier in life than is normally the case. The sufferer dies about 10 years after the onset of the condition. Some post-mortem examinations have revealed deposits (plaques) which contain aluminium within the brain. The interesting feature is that the aluminium is at the centre of focus of the plaques which implies that aluminium is the part of the mechanism that initiates plaque formation. It is these deposits that impede nerve-to-nerve communications which then induce the dementia.

Other surveys have shown a link between high incidences of dementia and low levels of calcium in the diet or drinking water. In one study, 30% of patients admitted to hospital with broken bones caused by calcium deficiency were also found to suffer from dementia. This linkage has yet to be firmly established but removal of aluminium from water supplies, diets and cooking utensils is a sensible precaution. In parts of Scandinavia, it is now standard practice to remove aluminium from water used for kidney dialysis treatment and for making up baby feeds.

Further reading

Beilke, S. and Elshout, A.J. (eds) *Acid Deposition*. D. Reidel, 1983, Dordrecht, The Netherlands.

Chadwick, M.J. and Hutton, M. *Acid Depositions in Europe*. Stockholm Environment Institute, 1991, York.

Cresser, M. and Edwards, A. *Acidification of Freshwaters*. Cambridge University Press, 1987, Cambridge.

Drabloes, D. and Tollan, A. (eds) *International Conference on the Ecological Impact of Acid Precipitation*. Sanderfjord, 1980, Oslo-Aas.

Duensing, E.E. and Duensing, L.B. *Environmental Effects of Acid Precipitation*. Vance Bibliographies, 1980, Monticello, Illinois.

Howells, G. Acid waters – the effects of low pH and acid associated factors on fisheries. *Advances in Applied Biology* **9**, 1983, 143–255.

Hutchinson, T.C. and Havas, M. (eds) *Effects of Acid Precipitation on Terrestrial Ecosystems.* Plenum Press, 1980, New York.

Last, F.T. and Watling, R. (eds) *Acidic Deposition: Its Nature and Impacts.* Royal Society of Edinburgh, 1991, Edinburgh.

Legge, A.H. and Krupa, S.V. (eds) *Air Pollutants and their Effects on the Terrestrial Ecosystem.* Wiley Interscience, 1986, New York.

Legge, A.H. and Krupa, S.V. (eds) *Acidic Deposition: S and Nitrogen Oxides.* Lewis Publishers, 1990, Michigan.

Longhurst, J.W.S. (ed.) *Acid Deposition: Sources, Effects and Controls.* British Library, 1989, London.

Schneck, J.J. *Acid Rain: A Critical Perspective.* Tasa, 1981, Minneapolis.

Schrader, S., Greve, U. and Schönwald, H.R. (eds) *Acid Precipitation and Forest Damage Bibliography.* Bundesforschungsanstalt für Forst- und Holzwirtschaft, 1983, Hamburg.

The Watt Committee on Energy. *Acid Rain.* Report No. 14, 1984, London.

Chapter 6

OZONE, PAN AND PHOTOCHEMICAL SMOG

'And the winds and sunbeams with their convex gleams build up the
blue dome of air'
from 'The Cloud' by Percy Bysshe Shelley (1792–1822).

Formation and sources

Tropospheric ozone formation

In the turbulent troposphere below the stratosphere, the major source of
O_3 is the photolysis of NO_2 (Reaction 6.1) which produces monatomic
oxygen (O). This then forms O_3 by reacting with molecular oxygen
(Reaction 6.2). Hydrocarbons, aldehydes and CO accelerate this initial
photolysis (Reaction 6.1) by increasing the rate of NO oxidation by
peroxy ($RO_2\cdot$) radicals (e.g. Reaction 6.3) and, at the same time, produce
highly reactive hydroxyl ($\cdot OH$) radicals. The various interrelationships

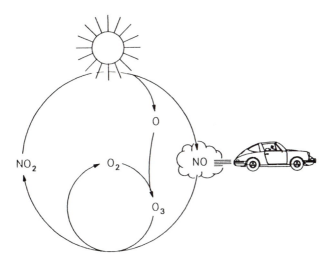

Fig. 6.1. Simplified diagram to explain the formation of O_3 from nitrogen oxides
in sunlight. The additional participation of unburnt HCs has been omitted, but
basically they prevent NO_2 returning as NO and increase the amount of O and
O_3 formed.

between the nitrogen oxides and O_3 are shown in detail in Fig. 3.5 and summarized in Fig. 6.1.

$$NO_2 + light \Rightarrow NO + O \tag{6.1}$$

$$O + O_2 + M^1 \Rightarrow O_3 + M \tag{6.2}$$

$$HO_2{}^\bullet + NO \Rightarrow NO_2 + {}^\bullet OH \tag{6.3}$$

Unburnt hydrocarbons

Evaporation of solvents and incomplete combustion of fuels release a wide range of hydrocarbons into the atmosphere. Analyses of ambient air samples at various times have shown the presence of over 600 different atmospheric hydrocarbons (HCs) which are mainly emitted north of 30°N. These include acetylene, benzene, butanes, ethane, hexanes, pentanes, propane and toluene – all of which are characteristic of homogenic emissions.

The most abundant HC in the atmosphere is the greenhouse gas, methane (CH_4 – see Chapter 8) which is released from decaying vegetation, as well as from industrial and domestic sources, in amounts ranging from 1 to 100 μl l^{-1}. Photolytic mechanisms exist for the complete oxidation of CH_4 to CO_2 (Fig. 6.2) as well as for the breakdown of other organic compounds. Certain HCs found in the atmosphere also act as plant growth regulators (ethylene or ethene, CH_2CH_4) while others are harmful to humans. Benzene, for example, induces cancer of the blood system following prolonged exposure and is a well-recognized hazard (see Chapter 9) for employees at roadside service stations.

Photochemical smog formation

The greatest global problem with unsaturated HCs is their ability to promote the formation of photochemical smog in the presence of nitrogen oxides, strong sunlight and stable meteorological conditions. The reaction chains involved are long and complicated because one reaction involving a free radical generates another. This, in turn, reacts to form a third, etc. Moreover, there are many starting points for these chain reactions because of the wide variety of HCs released to the atmosphere. Aldehydes and ketones also produce free radicals in strong light (Reaction 6.4) such as $RO_2{}^\bullet$ radicals (Reaction 6.5) while O_3 itself

---

[1] See Reactions 2.2 and 2.7.

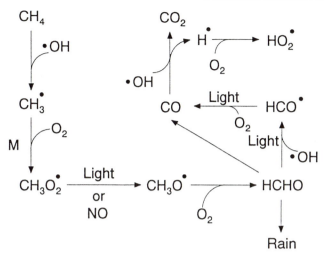

Fig. 6.2. Cascade of reactions that cause the oxidation of CH_4 to CO_2 typifying an atmospheric series of free radical-based reactions involving light, oxygen and ˙OH radicals. The last oxidation of CO to CO_2 is a very important reaction in the atmosphere.

may attack unsaturated HCs to produce similar free radicals including reactive aldehydes (Reactions 6.4 and 6.6). Many free radicals like these generated in strong sunlight are responsible for the eye irritation of those exposed to photochemical smog.

$$R - CHO + light \Rightarrow R^{\boldsymbol{\cdot}} + HCO^{\boldsymbol{\cdot}} \qquad (6.4)$$

$$R^{\boldsymbol{\cdot}} + O_2 \Rightarrow RO_2^{\boldsymbol{\cdot}} \qquad (6.5)$$

$$O_3 + R - CH{=}CH - R' \Rightarrow R - CHO + R'O^{\boldsymbol{\cdot}} + HCO^{\boldsymbol{\cdot}} \qquad (6.6)$$

$$R'O_2^{\boldsymbol{\cdot}} + RO^{\boldsymbol{\cdot}} \Rightarrow R'OR + O_2 \qquad (6.7)$$

$$R - CO.O_2^{\boldsymbol{\cdot}} + NO_2 + M^1 \Rightarrow R - CO.O_2{}^{\boldsymbol{\cdot}}NO_2 + M \qquad (6.8)$$

There are only a few reactions that terminate chain reactions. Rarely are they stopped by a double free radical collision (Reaction 6.7) – far commoner are reactions between $RO_2^{\boldsymbol{\cdot}}$ radicals and nitrogen oxides to form peroxyacyl nitrates (PANs, Reaction 6.8). PANs are highly reactive towards sensitive surfaces, such as eyes or delicate plant tissues, and the larger the carbon component of PANs the more toxic it is. Some eye irritants last longer than those that cause plant damage and involve other contaminants like formaldehyde (HCHO, Fig. 6.2) and acrolein.

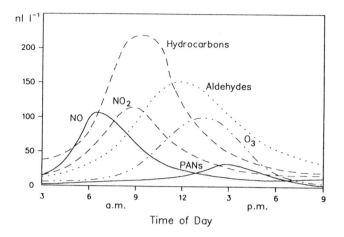

Fig. 6.3. Development and decay of components in a typical photochemical smog over a large conurbation during a bright, warm working day with no wind.

As a rough approximation, one molecule of PAN is produced for every 50 O_3 molecules.

Photochemical smog episodes have a distinctive daily rhythm (Fig. 6.3). In the absence of wind around dawn, when urban activities are low, levels of CO, NO and HCs increase as morning traffic builds up. Amounts of NO_2 then rise to a maximum about $1\frac{1}{2}$ hours after the peak of NO when the sun is reasonably high. The later appearance of aldehydes and then O_3 follows the disappearance of NO so that concentrations of O_3 are highest shortly after noon, just after the aldehyde peak has passed. This then allows other free radical chain reactions to form PANs and other irritants (Fig. 6.3). In late afternoon, returning traffic then generates more NO (and NO_2) which immediately scavenges most of the O_3 and reactive aldehydes. This means that levels of nitrogen oxides do not follow their distinctive biphasic daily pattern normally found in windy and often variable, cloud-covered conditions (see Chapter 1, especially Fig. 1.4).

The characteristic blue-brown haze over heavily populated cities, often associated with photochemical smog, consists mainly of unburnt HCs in an advanced state of oxidation. Typically, measurements of total oxidizing capability arising from O_3 and peroxides are highest in mid-afternoon and by midnight have almost disappeared. Conditions necessary for the formation of 'Los Angeles'-type smog exist throughout the world. The main requirements are roughly equal atmospheric levels of nitrogen oxides and HCs. Under conditions of bright sunlight and still air, most of the O_3 and PANs currently arise from NO and unburnt HCs emitted by mobile rather than static sources. Consequently, legislation to control exhaust gases from vehicles is more likely to reduce photochemically generated pollution in those areas susceptible to photochemical smog formation. Cost–benefit analyses have shown that restrictions on mobile sources are cheaper alternatives to retrofitting of pollution controls to existing power stations in those countries which have more sunlight, higher temperatures, and less wind.

Photochemical smog is now a worldwide problem, especially in cities of lower latitudes which have high rates of population growth and rapid industrialization. Currently, those most affected are Mexico City and Baghdad. Levels of O_3 in these and similar places are often over 100 nl l^{-1} for considerable periods of time. Moreover, tropospheric O_3 is no longer just confined to summer. Significant wintertime levels are now being experienced in cities like Madrid and Athens during periods when stable, high pressure, sunny conditions occur. Tropospheric O_3 produced over populous areas spreads out in the manner of low level clouds. Plate 7 shows northern America monitored by satellite over a period of 4 days when O_3 extended from Los Angeles to New York.

Damage mechanisms

Deterioration of materials

Ozone is such a reactive gas towards organic molecules that it is worth examining the consequences in detail. Any double bond in a HC is highly sensitive to chain-breaking and crosslinking reactions initiated by O_3. The mechanism of chain-cutting shown in Fig. 6.4 gives rise to $RO_2^{\bullet}$ radicals which then generate more free radicals in the same way that photochemical smog chain reactions are propagated (e.g. Reactions 6.6 and 6.7). The only difference is that they occur on the surface of materials and cause loss of tensile strength.

Natural polymers like rubber, cotton, cellulose, or leather, as well as paints, elastomers (used in tyre manufacture), plastics, nylon and fabric dyes, are all degraded by O_3. Only when the double bonds in their structures are protected by adjacent groups can attack be resisted. The electronegative chlorine atom adjacent to the double bond in neoprene is a good example of this type of protection. Similar protective mechanisms have been built into recent polymers but sacrificial autoxidants can also be incorporated. The annual global cost of this type of damage to materials is huge and constitutes just over 30% of that caused by all forms of atmospheric pollution to non-living materials.

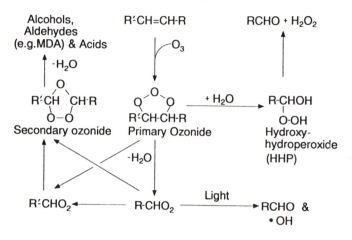

Fig. 6.4. Basic sequence of reactions associated with the process of ozonolysis. The main points to note are that the presence or absence of H_2O affects the possible reactions and that, in a biological context, free radicals are not involved in aqueous phase reactions.

Ozonolysis or peroxidation?

Most components of biological tissues are in close proximity to water. This changes the characteristics of O_3 attack upon unsaturated bonds in carbon compounds. Instead of forming free radicals and secondary ozonides, hydrated forms of O_3 form H_2O_2, hydroxyhydroperoxides (HHPs), and reactive aldehydes (Fig. 6.4) by a process known as ozonolysis. If the compound under attack by O_3 has more than one double bond (e.g. the fatty acids of lipids) then malondialdehyde ($OHC.CH_2.CHO$) is also formed (Reaction 6.9). The same compound may also be produced from similar substrates by a different series of reactions known as lipid peroxidation. In the minds of some, ozonolysis and lipid peroxidation are synonymous, but in fact they are quite distinct processes. The former creates H_2O_2, does not bring double bonds into conjugation with each other (i.e. produce alternative double and single $C-C$ bonds) and, in aqueous phases, does not involve free radicals. Lipid peroxidation, on the other hand, produces conjugated products (alternate single and double bonds), involves initial attack by free radicals rather than by O_3, and does not form H_2O_2 (Fig. 6.5).

Plate 1 Ranked smooth meadow grass (*Poa pratensis* cv. Monopoly) tillers grown to harvest in SO_2-polluted air for 20 weeks, illustrating the large genetic variation in response to pollution even in one cultivar of a species. Similar results were obtained with NO_2 and SO_2 + NO_2 fumigations.

Plate 2 Hydroponically grown lettuces (cv. Ambassador) grown for 21 days in clean air (**right**) or 2 µl l^{-1} NO (**left;** typical of CO_2-enriched commercial glasshouses heated by propane). The front plants have minimal nitrate (0.5 mM) and the rear ones adequate nitrate (10 mM). This experiment demonstrates that NO inhibits growth but when nitrate supply to the roots is limited, NO does not usually become an alternative source of nitrogen.

Plate 3

Plate 4

Plate 5

Plate 6

Plates 3–6 Examples of visible injury due to SO_2 on leaves of alfalfa (*Medicago sativa* cv. Du Puits, **Plate 3**); to O_3 on needles of ponderosa pine (*Pinus ponderosa*, **Plate 4**); to fluoride on leaves of gladioli (*Gladiolus gandavensis* cv. Snow Princess, **Plate 5**); and injury due to O_3 on an O_3-sensitive tobacco leaf (*Nicotiana tabacum* cv. Bel W3, **Plate 6**)

Plate 7 Two Nimbus-7 satellite pictures taken on successive days (5th and 6th of April, 1982) showing the UV-absorbing regions over northern America. The purple and green areas (fading to white, tan and brown) show high levels of tropospheric O_3 extending across the USA and Canada. In southern Mexico, the black area devoid of O_3 is due to the scavenging effect of SO_2 in the volcanic cloud emitted from the eruption of El Chicon.

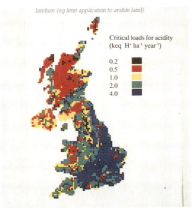

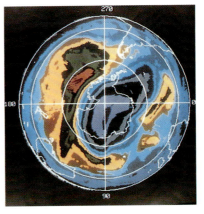

Plate 8 Critical load map of acidic deposition of soils in the UK, based on soil mineralogy, weathering rates and any moderating effects due to liming.

Plate 9 Satellite picture of the Antarctic region taken in springtime, showing very low levels of UV-absorbing O_3 over the pole (blue-black) and much higher O_3 levels in the southern Pacific Ocean (red-green).

Plate 10

Plate 11

Plate 12

Plate 13

Plates 10–13 A Norway spruce tree (*Picea abies,* **Plate 10**) showing extensive needle loss, especially just below the crown, and streamer-like hanging of the lower branches. **Plate 11** – two branches of Norway spruce collected on the lower slopes of the Jungfrau, Switzerland, one of which shows typical upper surface yellowing of older needles, characteristic of either natural or ammonium-induced magnesium deficiency. **Plate 12** – Sitka spruce tree (on left, clean air control on right) showing yellowing symptoms of recent forest decline in springtime, after 3 previous summer fumigations with episodes of O_3. **Plate 13** – tobacco plants after experimental exposure to UV-B radiation (plant on left) or no UV-B (right). The UV-B-treated leaves are highly reflective and curl upwards.

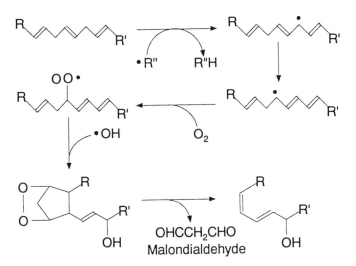

Fig. 6.5. Reaction sequence to illustrate the basic features of lipid peroxidation. Free radicals are involved both at the start and during the various reactions. Also note that one of the products, malondialdehyde, is not unique to peroxidation but may be also produced during ozonolysis (Fig. 6.4).

These distinctions are vital for an appreciation of which series of reactions and reactive species actually attack the proteins or lipids of biological membranes. Production of malondialdehyde is not evidence for one particular process because it is produced by both. Experimental O_3-treatments have shown that free radicals from lipid peroxidation are unlikely in biological materials at ambient temperatures, but H_2O_2 formed by ozonolysis is often detected. O_3 also forms ˙OH radicals (Reaction 6.10) in alkaline solutions or by reacting with H_2O_2 (Reaction 6.11) but, in living systems, the latter is quickly removed by catalase (Reaction 6.12 – see later).

$$O_3 + R-CH=CH.CH_2.CH=CH-R'$$
$$\Rightarrow R-CHO + OHC.CH_2.CHO + R'-CHO \quad (6.9)$$

$$O_3 + H_2O \Rightarrow 2\text{˙OH} + O_2 \quad (6.10)$$

$$H_2O_2 + O_3 \Rightarrow HO_2\text{˙} + \text{˙OH} + O_2 \quad (6.11)$$

$$2H_2O_2 \overset{\text{catalase}}{\Rightarrow} 2H_2O + O_2 \quad (6.12)$$

$$NO_2 + R-CH=CH.CH_2-R' \Rightarrow HNO_2 + R-CH=CH.C\text{˙}H-R' \quad (6.13)$$

There is less misunderstanding about the attack of low levels of nitrogen oxides on similar unsaturated compounds. These involve the formation of free radicals, as well as nitrous acid (Reaction 6.13), and show great similarity to those of lipid peroxidation. If this was the mechanism which gives rise to toxicity in both plants and animals then both O_3 and nitrogen oxides should be equally toxic and produce similar types of damage. However, O_3 is more harmful than nitrogen oxides, which indicates a different type of attack by O_3 (ozonolysis) to that by the nitrogen oxides (lipid peroxidation).

Proteins are more sensitive than lipids

Three amino acids (cysteine, methionine and tryptophan) are especially sensitive to attack by O_3 in both aqueous solutions and biological tissues. The sulphydryl groups ($-SH$) of the first two amino acids are oxidized to disulphide bridges ($-S-S-$) or to sulphoxides ($>S=O$) while the pyrrol ring of tryptophan is opened up to form *N*-formylkynurenine (Fig. 6.6).

However, if O_3 reacts with the same amino acids while they form part of the proteins which carry out critical functions within the cell then the impact is greater. Such disturbances are especially damaging if one or more of these amino acids affect the secondary and tertiary shapes of a protein. Changes in spatial orientation of proteins are critical if, for example, they form part of the reaction centre of an enzyme. Some enzymes certainly show clear evidence of this type of attack. Furthermore, there is little evidence that once damaged by O_3, the original activity or shape of an enzyme is regained.

Proteins in plants, microbes and animals most likely to be exposed to O_3 attack are those partially embedded within cell membranes. Indeed, these proteins often show changes due to O_3 well before the lipids inside the membrane beneath. Changes to membranes due to attack on both proteins and lipids, however, induce pronounced changes in cell permeability, etc. The consequences of this are considered later.

Reactions of PANs

Less research has been carried out to follow the mechanisms of damage by PANs. This is largely because they are very difficult to generate in laboratory conditions. PANs attack those amino acids which are also sensitive to O_3 (e.g. methionine to methionine sulphoxide or cysteine to cystine; see Fig. 6.6). This means that changes to the thiol groups of proteins again involve changes in conformation and loss of enzyme

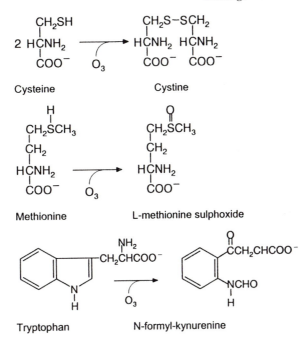

Fig. 6.6. Some groups in amino acids are especially sensitive to O_3. These include the sulphydryl groups of cysteine which form disulphide bridges or the sulphur atoms of methionine which are oxidized to sulphoxide. Tryptophan is also very sensitive to opening of the pyrrol ring. All these reactions take place either with the free amino acids or when they are already incorporated into enzymes and other proteins. In the latter case, the damage caused is greater.

activity. Fortunately, penetration of PANs into proteins is poor and the half-life of most PANs is short (about 7 minutes at pH 7). Moreover, some key biological electron donors like NADH and NADPH (see Chapter 2) are only oxidized by PANs and not destroyed as is the case with O_3.

On the other hand, PANs absorb more of the protective natural antioxidants than O_3. Normally, O_3 converts two molecules of reduced glutathione (a tripeptide called gamma-glutamyl-cysteinyl-glycine, GSH) to a molecule of oxidized glutathione (GSSG, Reaction 6.14). PANs, however, also react with a third molecule of reduced GSH (Reaction 6.15) which cannot be easily replaced by natural GSH regenerating mechanisms (glutathione reductase, etc. – see later). This reduces the pool size of natural antioxidants and, as a consequence, more damage follows.

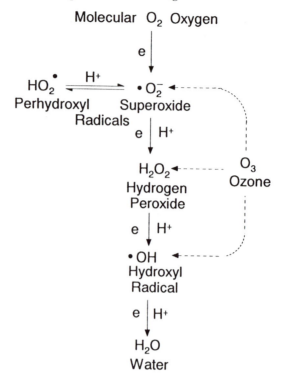

Fig. 6.7. Reduction of oxygen through its various forms. Note that O_3 forms a number of different reactive states and that the $^\bullet O_2^-$ radical is in equilibrium with the $HO_2^\bullet$ radical in a one-to-one relationship at pH 4.7.

$$2G-SH^1 + O_3 \Rightarrow G-S-S-G + H_2O_2 + \tfrac{1}{2}O_2 \qquad (6.14)$$

$$3G-SH^1 + CH_3CO.O_2NO_2 \Rightarrow G-S-S-G + G-S.CO.CH_3$$
$$+ H_2O + HNO_3 \quad (6.15)$$

Natural antioxidants

Univalent reduction of molecular oxygen produces superoxide ($^\bullet O_2^-$) and $^\bullet OH$ radicals, H_2O_2, and H_2O (Fig. 6.7) as well as O_3. $^\bullet O_2^-$ and H_2O_2 are not themselves sufficiently reactive in aqueous solution to initiate lipid peroxidation but the perhydroxyl radical ($HO_2^\bullet$) can. The pK_a of the equilibrium between $^\bullet O_2^-$ and $^\bullet OH$ (Reaction 6.16) is 4.8 which means that, in well-buffered cell contents (around pH 7), only a few $HO_2^\bullet$ radicals exist. In cell vacuoles, however, where acidity levels are

[1] G = glutathione, a tripeptide.

higher, their formation is more likely. Furthermore, unbuffered extra-cellular fluids exposed to both SO_2 and NO_2 (see Chapter 11) also form $HO_2^{\bullet}$ at the expense of ${}^{\bullet}O_2^{-}$.

$$H^+ + {}^{\bullet}O_2^{-} \Leftrightarrow HO_2^{\bullet} \tag{6.16}$$

In an aqueous environment, ${}^{\bullet}O_2^{-}$ is relatively unreactive but, once in the hydrophobic regions of the fatty acid side chains within membranes, it oxidizes α-tocopherol (vitamin E) and removes the most important non-aqueous antioxidant mechanism in a cell. ${}^{\bullet}O_2^{-}$ and $HO_2^{\bullet}$ radicals, however, may be converted via H_2O_2 to ${}^{\bullet}OH$ radicals (Reactions 6.17–6.19) which then react with virtually all biological molecules immediately. This means they do not diffuse any distance before reacting.

$$ {}^{\bullet}O_2^{-} + Fe^{3+} \Rightarrow Fe^{2+} + O_2 \tag{6.17}$$

$$\text{hypoxanthine}^2 \text{ [or xanthine}^2] + O_2 \Rightarrow$$
$$\text{xanthine}^2 \text{ [or urate}^2] + H_2O_2 \tag{6.18}$$

$$Fe^{2+} + H_2O_2 \Rightarrow Fe^{3+} + OH^- + {\cdot}OH \tag{6.19}$$

$$2{}^{\bullet}O_2^{-} + 2H^+ \Rightarrow H_2O_2 + O_2 \tag{6.20}$$

Many mechanisms also release H_2O_2 and these often form part of the antioxidant defences of biological tissues. Some are specific to certain species but a great number occur in all aerobic (oxygen-requiring) organisms. Superoxide dismutases (SODs), for example, are metallo-proteins which convert ${}^{\bullet}O_2^{-}$ into H_2O_2 (Reaction 6.20). SODs are of three basic types – a copper–zinc protein common to vertebrates, higher plants and fungi, a manganese enzyme found in all animals and plant mitochondria as well as bacteria, and an iron-containing SOD only found in certain prokaryotic organisms (i.e. those lacking internal membrane-bound compartments). To operate properly, SODs require the cooperation of another widely distributed enzyme (catalase) to break down the H_2O_2 (Reaction 6.12) which would otherwise form ${}^{\bullet}OH$ radicals (Reaction 6.19).

At least three other lines of antioxidant defence are common to most organisms – all of which are semi-sacrificial. Glutathione peroxidase, for example, removes H_2O_2 and other organic peroxides from aqueous parts of the cells by catalysing the conversion of two molecules of glutathione (GSH) to one oxidized molecule of glutathione (GSSG – Reaction 6.21). Reduced glutathione is then regenerated from GSSG by another enzyme, glutathione reductase (Reaction 6.22).

---

[2] Purine bases.

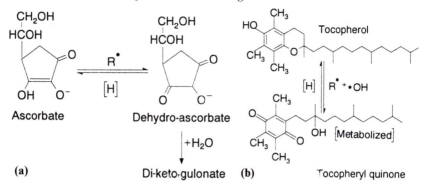

Fig. 6.8. Two of the most important antioxidant systems in most biological systems are (a) ascorbate (Vitamin C) and (b) α-tocopherol (Vitamin E). In both systems, free radicals are scavenged and the oxidized products are partially recovered by separate reduction processes. Ascorbate is an effective antioxidant in aqueous environments while α-tocopherol protects sensitive hydrophobic compounds such as lipids within membranes.

$$2G-SH^1 + ROOH \Rightarrow G-S-S-G + H_2O + ROH \quad (6.21)$$

$$G-S-S-G^1 + 2NADPH \xrightarrow{\text{glutathione reductase}} 2G-SH + 2NADP^+ \quad (6.22)$$

Vitamin E (α-tocopherol) is formed by plants but is also required in the diet of humans. In both cases, it is used as a semi-sacrificial antioxidant in the lipids of cell membranes. It is oxidized first to a semiquinol and then to tocopheryl quinone by $^{\bullet}O_2^-$, H_2O_2 or $^{\bullet}OH$ (Fig. 6.8). Some tocopheryl quinone may be reduced back to α-tocopherol for reuse, but overall losses are considerable and must be replaced by fresh synthesis in plants or through the diet by animals.

Vitamin C (ascorbate) is another antioxidant made by plants, taken up in the diet of humans, and used in a similar scavenging role in both animals and plants. This carbohydrate reacts with $^{\bullet}O_2^-$, H_2O_2 and $^{\bullet}OH$ radicals to form dehydroascorbate when catalysed by two enzymes, ascorbate oxidase or ascorbate peroxidase (Fig. 6.8). Enzymes to reduce dehydroascorbate back to ascorbate are also present, but inevitably some ascorbate is lost as the sugar ring opens to form 2,3-diketo-gulonate which is then metabolized away. Constant replenishment of ascorbate by recycling, or by fresh biosynthesis in plants and release from restricted reserves in animals, is continuously required to maintain high levels of this essential antioxidant in the aqueous parts of cells.

Other antioxidant systems exist, especially in plants. The phenolic components of plant cell walls and certain carotenoids (xanthophylls)

[1] G = glutathione, a tripeptide.

inside chloroplasts can absorb potential oxidants. In the xanthophyll cycle, zeaxanthin is epoxidized to antheraxanthin and then to violaxanthin. A separate enzyme system then removes the epoxy-groups to regenerate zeaxanthin.

Any consideration of attack upon different biological systems must, therefore, always take natural protective mechanisms into account. The nature of the oxidants and where they are in a cell also determines which antioxidant is consumed. Aqueous parts of cells favour the use of glutathione or ascorbate (and phenols in plants), while non-aqueous (or hydrophobic or lipophilic) parts of cells are more dependent on the protection offered by α-tocopherol and, in the case of plants, carotenoids as well.

Plant effects

Entry

Deposition velocities of O_3 and PANs to various surfaces (e.g. soil, water and vegetation) are higher than those for nitrogen oxides and about the same as those of SO_2 (Table 1.4). Still conditions during photochemical smog episodes provide some protection to vegetation because undisturbed boundary layers of still air around leaves offer additional resistance to the diffusion of O_3. Later, as winds move tropospheric O_3 around, damage to vegetation may be experienced many kilometres away as these leaf boundary layers are stripped away.

Access of O_3 (and PANs) into the leaf takes place through the stomata although waxy leaf surfaces (cuticles) may be partially degraded by O_3. Most plants have open stomata during the day. Consequently, O_3 damage is initiated in the light but cacti, etc., which only open their stomata at night to conserve water loss, have their period of maximum sensitivity to O_3 during periods of darkness.

Responses to O_3 are determined by a wide variety of environmental or genetic factors (see Table 6.1). Moist surfaces within the leaves (e.g. the extracellular fluid of mesophyll tissues) allow O_3 to dissolve and diffuse down a concentration gradient similar to that of CO_2. Solubilities, rates of decomposition, and the pH values of the various media all influence the amount of O_3 that is taken up. O_3 is approximately one-third as soluble as CO_2 but 100-fold less soluble than SO_2 (see Chapter 2).

Cellular changes and damage

Once in the air spaces within a leaf and close to the extracellular fluid, O_3 immediately forms other derivatives which show varying degrees of

Table 6.1   Factors that influence plant response to O_3 and PANs

Genetic	Environmental

Development (and senescence):
  Cell (including vigour and age)
  Tissue (including succulence)
  Leaf (including shape, hairs, waxes)
  Root
  Plant (including stomatal structure)
	Soil conditions:
	  Acidity
	  Temperature
	  Water stress
	  Nutrients (K^+, phosphate, nitrate)

Within and between species:
  Individuals
  Clones
  Varieties
  Cultivars
  Populations
	Climatic conditions:
	  Light
	  Humidity
	  Temperature
	  CO_2 levels
	  Other pollutants (PANs, HCs,
	    SO_2, NO, NO_2, HF, NH_3)
	  Duration of pollution episodes

	Biotic conditions:
	  Pathogens (insects, fungi, bacteria,
	    viruses)
	  Individual competition
	  Species competition
	  Pesticides
	  Mycorrhizal associations

reactivity. Indeed, it is unlikely that much unreacted O_3 passes further into cells. Studies at Lancaster and elsewhere have shown that O_3 reacts with unsaturated HCs, some of which are emitted by the cells beneath (ethene or isoprenoid compounds), to form primary ozonides and hydroxyhydroperoxides (HHPs, see Fig. 6.4). It is these HHPs that are then responsible for some of the harmful effects of O_3 which take place before the natural antioxidant systems within cells can counteract their effects.

Most studies of O_3 damage show that the plasma or cell membranes of plant cells (sometimes called the plasmalemma) suffer the most injury. This is characterized by changes in permeability and 'leakiness' of cell membranes to important cations like potassium. Internal membranes (e.g. organelle envelopes) are affected to a lesser extent as the toxic oxidants derived from O_3 are diluted and absorbed from the outside inwards.

While there is some evidence of cellular repair, sites of injury are unspecific. Heavy damage due to oxidants arising from O_3 is heralded by losses of chlorophyll, increases in leaf fluorescence (indicative of wasted light energy), and changes in adenylate (ATP, etc.) levels. More sensitive perturbations within plant cells due to O_3 show changes in the fluxes of various components across membranes, especially those of sugars, amino acids, water and potassium ions. Disturbed transport across membranes may be traced to a failure to control osmotic pressure and maintain effective electropotentials across membranes. Only when levels of O_3 are low, or only present for short periods, is subsequent recovery of normal transport capability possible.

Visible injury

Injury induced by HHPs, etc., arises from an inability to repair or compensate for altered membrane permeability. Initially confined to sensitive cells, injury extends to irreversible pathological damage characterized by bleaching of cells. Early symptoms of injury appear as 'water-soaked' areas just below the epidermis. If recovery is still possible these shiny water-soaked regions gradually disappear as the tissues recover their control of permeability.

Visible injury is normally confined to foliage but a wide variety of genetic and environmental factors influence the nature of visible injury (see Table 6.1). The usual type of damage is chlorotic flecking or stippling of the leaves between the veins (see Plate 6) but the nature of damage varies between species. Later, such injuries may become reddish or bronzed due to enhanced anthocyanin production or darken due to tannin formation. Plate 4 shows the characteristic yellow and white banding of conifer needles caused by repeated photochemical smog episodes, while the front cover of this book shows extensive bronzing of Ponderosa pines in the San Bernardino mountains of southern California (courtesy of Dr Mark Poth, US Forest Service, Riverside).

Peroxyacyl nitrate damage

Peroxyacyl nitrates (PANs) also enter a leaf through the stomata and, molecule for molecule, are more toxic to plants than HHPs, etc. Fortunately, they generally occur at much lower concentrations. Visible symptoms also differ with bronzing of the lower surfaces of leaves as a regular feature. Developing regions of leaves and needles are especially sensitive to PANs and this gives a 'stripe-like' appearance when a smog episode coincided with growth at that point (see Plate 4). On either side of these regions, the adjacent tissue remains healthy.

Targets of damage are the sulphydryl groups of protein and, to a lesser extent, the unsaturated double bonds of lipids. Furthermore, there must be illumination before, during and after exposure to PANs for damage to occur. The reasons for this are complex but sunlight initiates the formation of additional free radicals which then overwhelm the antioxidant mechanisms. Unrepaired areas of damage then deteriorate as the products of photooxidation build up which, in turn, start fresh and uncontrolled disturbances.

Health effects

Hazards indoors and at work

Problems due to O_3 pollution indoors are unappreciated by most people and the situation is not improving. In most situations, it is O_3 that seeps in from outside that causes most problems. Given average rates of air changes (4 h^{-1}), levels inside offices, etc., are 70% of those outside. This means that indoor O_3 levels are also higher in summer and lower in winter. In areas prone to photochemical smog, indoor levels of O_3 over 100 nl l^{-1} have been recorded.

Wide variations exist which are due to wide differences in rates of air changes (0.6–30 h^{-1}) and the internal furnishings. Deposition rates of O_3 to different common surfaces vary widely. Those to modern finishes like plastics and glass are very low (*ca.* 0.001 cm s^{-1} to new glass) but very high to fresh fabrics (0.109 cm s^{-1} for new cotton).

In the past, there was some appreciation by those working with specialized equipment such as X-ray machines, UV lamps, xenon arc lighting and other high voltage-generating equipment that a major hazard was the release of O_3 due to ionizing electrical discharges. Probably the most hazardous task involving O_3 is electric torch welding, which is often carried out in restricted spaces. In these circumstances, it has always been safe practice to ensure adequate ventilation to reduce the problem. Similarly, aircraft flying at very high altitude have to take into account the higher levels of O_3 in the lower stratosphere and to provide adequate protection for air travellers.

However, the advent of photocopiers and fax machines throughout many workplaces, often in places not designed for such equipment, has increased indoor O_3 levels. Repetitive noise in the office usually causes such machines to be semi-isolated to poorly ventilated corners or small side rooms and the problem is made worse. Recommended TLVs of O_3 during a 40 h working week of below 0.1 μl l^{-1} O_3 are still too high and levels indoors well below 0.05 μl l^{-1} are more desirable.

Hazards outside

Natural levels of O_3 outdoors are around 0.005 μl l^{-1} but lightning strikes or photochemical smog episodes may raise levels well above 0.1 μl l^{-1}. The USA standard of an hourly mean of 0.12 μl l^{-1} O_3 outdoors (not to be exceeded more than one day per year) is exceeded many times during summer months in many areas of the USA, Europe, and Japan.

High risk groups include children, the aged, individuals with pre-existing respiratory or immune-related diseases, and those with coronary heart disease. Moreover, it is the concentration of O_3 rather than the length of exposure that influences response. A short exposure to high concentration causes proportionally more harm than a long exposure to low levels of O_3. Similarly, intermittent exposures are worse than continuous exposures. This means that 6 hours of O_3 and clean air for the rest of the day may be worse than O_3-polluted air at a similar concentration for the whole day. If other pollutants are also present (e.g. nitrogen oxides) then the effects are more than additive, and are sometimes described as synergistic (see Chapter 11).

Short- and long-term exposures

Photochemical smog causes irritation to the eyes, nose, throat, and chest. Eye irritation is not caused by O_3 but by PANs and trace free radical HCs. However, little research has been done to distinguish the physiological responses to O_3 from those due to PANs, etc., although extensive studies have determined the conditions under which O_3, etc., cause tumours. Once again, this has mainly been done by using animals as model systems with all the attendant problems of translating observed animal responses into likely human consequences (see Chapters 2 and 3), and the difficulties of interpretation remain. Nevertheless, it has been established that atmospheric levels of O_3, at least, are not carcinogenic.

There is also agreement that O_3 is a powerful oxidant which may injure the bronchiolar and alveolar walls of the lungs. Surface epithelial cells of the airways are damaged by O_3 and, afterwards, these are replaced by thick cuboidal cells with few or small cilia (cell hairs). Additional changes to the epithelial (surface lining) cells during O_3 injury, apart from loss of cilia, include cytoplasmic vacuolation (formation of internal cellular spaces) and condensation of abnormal mitochondria. Similar cellular damage may be caused by NO_2 but at 20 times the concentration of O_3. Initial damage is often accompanied by an accumulation of water within affected tissues (oedema) and together these symptoms may give rise to acute inflammatory responses. These only occur at high levels of O_3 exposure and diminish when O_3 levels are reduced.

Effects of O_3 are also made worse by exercise. It is now recommended that athletic activities should cease when levels of O_3 are high and that groups at risk (e.g. children, who spend proportionately more time outdoors than any other group in the population) are moved indoors.

Over long periods of exposure, replacement of lung epithelial cells

Table 6.2   Known cross-tolerance relationships between air pollutants for humans

Primary agent	Secondary agents which induce tolerance to the primary agent
H_2S	O_3, $COCl_2$
NO_2	CCl_3NO_2, O_3, $COCl_2$, thiourea
O_3	CCl_3NO_2, H_2S, ketene, NO_2, $NOCl$, $COCl_2$, thiourea

with cuboidal cells may be considerable. There is also an increase in the number of macrophages (certain white blood corpuscles), fibrous elements, and mucous-secreting cells, as well as a thickening of the walls of the major airways. These changes exacerbate a number of clinical problems such as bronchitis and emphysema. Those suffering from these complaints form the group most at risk to atmospheric O_3. Children, especially those who live in degraded environments which are also prone to photochemical smog, now form a high proportion of those suffering from bronchitis.

Following initial exposure, however, enhanced tolerance to subsequent O_3 exposures has been reported. Higher concentrations of O_3 have been found to give rats more protection against re-injury than lower levels of O_3 when subsequently re-exposed to double the initial concentrations. Similarly, when biochemical changes in the blood of California residents were compared to those of Canadians exposed to similar concentrations of O_3, a number of indicator tests showed that some tolerance had already been acquired by residents of California.

Epidemiological studies of enhanced respiratory infections and reduced lung function in populations exposed to photochemical smogs are hampered by the fact that such surveys are dealing with complex and changing mixtures of O_3, nitrogen oxides, PANs, and other components of smog. Shorter-term dose–response studies have indicated that, although nitrogen oxides and O_3 cause similar adverse effects on lungs, more than five times the level of NO_2 are required to elicit similar effects to those of O_3 – very like the plant situation.

Cross-tolerance mechanisms have also been extensively studied and O_3 has been found to initiate a variety of cross-tolerance reactions. A number of these instances when one toxic substance induces the acquisition of tolerance to another are known and some of these are shown in Table 6.2.

Biochemical and physiological changes

Animals also have a wide range of protective antioxidant scavenging systems (see the section above on Reactions of PANs). Exposure to O_3

causes increases in levels of SODs, glutathione peroxidase, glutathione reductase, disulphide reductase, and non-protein sulphydryl groups (mainly glutathione) within lung tissues. These, in turn, partially account for some acquired tolerance.

Levels and rates of consumption of ascorbate have not been intensively studied, but animal tissues which are adequately supplied with vitamin E (α-tocopherol) are much less sensitive to O_3 than those deficient in this protective compound. Increased production of malondialdehyde, indicative of both ozonolytic and lipid peroxidation, also occurs. However, there is no reason to believe why O_3 attack in animals is any different from that in plants. Within the moist mucosal fluid layers covering the airway cells, O_3 probably reacts with unsaturated HCs to form HHPs and reactive aldehydes, etc. (Fig. 6.4) which then go on to damage plasma membranes.

Disturbances of mitochondrial structure, reductions in lung tidal volume, and increases in rates of respiration are all known to be induced by breathing O_3-polluted air. These are accompanied by swelling of cells, increased rates of metabolic enzymes, enhanced oxygen consumption, and uncoupling (inefficient formation of ATP). However, these disturbances by O_3 within mitochondria are probably secondary. Oxidant-induced damage to the permeabilities of cell membranes precede them – in a very similar way to those changes that occur in sensitive plant tissues.

Maintaining the sterility of lung tissues is a major function of alveolar macrophages. These white blood cells, derived from the bone marrow, surround invading particles (including bacteria) with a protective vacuole called a phagosome, and enzymes (e.g. acid phosphatase and lysozyme) are also placed in this phagosome by other organelles called lysosomes. During the following digestion, the vacuolar membrane protects the rest of the macrophage from these powerful breakdown enzymes as they work upon and finally destroy the incapacitated invaders. Normally, the hydrolytic enzymes of the macrophages are carefully packed away in lysosomes to avoid harm being done to the rest of the macrophage. However, the products arising from O_3 are capable of causing increased fragility and partial lysis of lysosomes. Usually, it is only upon death of a cell that the hydrolytic enzymes of the lysosomes are released into the rest of the cell to bring about self-digestion. The products of O_3 (HHPs, etc.), therefore, have the effect of initiating this possibility before the normal lifespan of a cell is complete.

Exposure of lung tissues to O_3 also appears to induce macrophages to congregate just as though a bacterial invasion had already been experienced, even though no actual infection has started. More macrophages are then drawn in and they also become partially disabled, depending on the degree of exposure to atmospheric O_3. At the same time, their original job of fighting bacterial infection is partially suspended so that pneumonia and other respiratory illnesses take hold more easily. The consequences of O_3 are therefore made worse by changes in normal cellular activities which exist for other purposes. This gives some insight into how low levels of atmospheric O_3 bring about significant changes in lung function which create special difficulties for those already suffering from bronchitis and emphysema.

Cellular models

Erythrocytes (red blood cells, or RBCs) carry vast amounts of O_2 around the body. To do this, the cell membranes are freely permeable to non-reactive dissolved gases. O_2 is actually very soluble in the hydrophobic (water-repelling) regions of membranes, but O_3 is 10 times more soluble than O_2 in aqueous blood plasma. Consequently, human RBCs have been much used as model cells to study effects of O_3 on cell membranes, because they have the advantage that they do not synthesize protein after they have lost their nuclei and only metabolize sugars using glycolytic and associated pathways. As a consequence, inherent adaptive biochemical processes which might occur during experimental exposure are not a problem. Studies with RBCs have revealed numerous sites of sensitivity to O_3 and H_2O_2. For example, changes in membrane fragility, depressed levels of antioxidants such as glutathione, and changes in enzymic activity have all been detected.

The mechanisms of O_3 attack on individual proteins of RBCs have also been examined. In glycophorin, a transmembrane protein, amino acids numbered 72 to 92 are embedded within membranes and do not contain cysteine or tryptophan. The methionine at number 8, however, is on the outside of membranes while that at position 82 inside. When membranes are exposed to O_3, methionine 8 is oxidized to methionine sulphoxide (Fig. 6.6) but methionine 82 is protected. Changes like this explain why immune responses, which depend on exposed surface proteins, are altered by O_3.

Experiments with RBCs have also demonstrated that levels of O_3 have to be higher for effects on lipids as well as proteins to occur. Addition of a lysophospholipid to RBCs causes them to break open or lyse – a process very similar to that which often occurs during a snake bite or insect sting – but similar exposures of RBCs to O_3 do not cause lysis. However, exposure of phospholipids to O_3 in the absence of RBCs causes the fatty acids in the phospholipids to undergo ozonolysis (i.e. no free radicals are formed) which then behave as if they were lysophospholipids and are able to lyse or break open RBCs in the absence of O_3. This means that O_3 cannot generate ozonized phospholipid while it is buried within the cell membrane yet it can change exposed sulphydryl groups on the outside of cell membranes (from the glycophorin experiments). Consequently, attack of O_3 is on the outward-facing proteins rather than on the lipids inside membranes. It is only after this primary damage in proteins has changed membrane function (i.e. increased permeability) that secondary disturbance of lipids takes place.

Mutagenesis

Ultraviolet light and O_3 are commonly used as alternatives to hypochlorite to kill microorganisms in swimming pools. The O_3 both oxidizes the sulphydryl groups and ozonates the fatty acids of bacterial cell walls, but it also interacts with UV light to cause genetic damage which involves rupture of DNA strands and damage to the DNA repair mechanisms. The chromosomal disruption is initiated by UV light but the extra presence of O_3 then inhibits the DNA recovery mechanisms that would normally follow.

Many years ago, extensive animal trials evaluated the mutagenic risks to humans of atmospheric O_3, PANs, etc. Little evidence was produced to support such a possibility, but similar studies using irradiated smog from car exhausts produced a marked decrease in number of mice per litter, frequency of litters per mother, and survival rates of infant mice. The elevated mortality rate was traced to alterations in the genetic composition of the sperm which must have been caused by components other than O_3 in the smog.

Over the past two decades, studies of the effects of active agents of photochemical smog, other than O_3, on animals and plants have been less common and yet the need to evaluate them properly has never been so great. The long-term health hazards of many other active ingredients of photochemical smog and trace HCs, other than those that also occur in tobacco smoke, have still to be evaluated.

Further reading

Adams, R.M., Hamilton, S.A. and McCarl, B.A. *The Economic Effects of Ozone on Agriculture.* US Environmental Protection Agency, 1984, Corvallis, Oregon.

Berglund, R.L. (ed.) *Tropospheric Ozone and the Environment.* Air and Waste Management Association, 1992, Pittsburgh, Pennsylvania.

Grennfelt, P. (ed.) *Ozone – The Evaluation and Assessment of the Effects of Photochemical Oxidants on Human Health, Agricultural Crops, Forestry, Materials and Visibility.* Swedish Environment Research Institute, 1984, Gothenburg.

Guderian, R. and Rabe, R. *Photochemical Oxidants – Formation, Control, Effects on Man, Animals and Plants.* Springer-Verlag, 1983, Berlin.

Guderian, R. (ed.) *Air Pollution by Photochemical Oxidants.* Springer-Verlag, 1985, New York.

Halliwell, B. and Gutteridge, J.M.C. *Free Radicals in Biology and Medicine.* Clarendon Press, 1985, Oxford.

Levitt, J. *Responses of Plants to Environmental Stresses.* Academic Press, 1972, New York.

Pell, E.J. and Steffen, K.L. (eds) *Active Oxygen/Oxidative Stress and Plant Metabolism*, vol. 6, *Current Topics in Plant Physiology.* American Society of Plant Physiology, 1992, Rockville, Maryland.

Chapter 7

STRATOSPHERIC OZONE DEPLETION AND ENHANCED UV-B RADIATION

'Atoms or systems into ruin hurl'd, and now a bubble burst, and now a world.'
from 'An Essay on Man' (1733) by Alexander Pope (1688–1744).

Ultraviolet radiation

Although largely dependent upon wavelength, certain frequencies of solar radiation penetrate the entire atmosphere and reach the surface of the Earth (Fig. 7.1). Very long radio waves do not get much further down than 50 km above sea level but shorter radio waves (1 cm to 10 m)

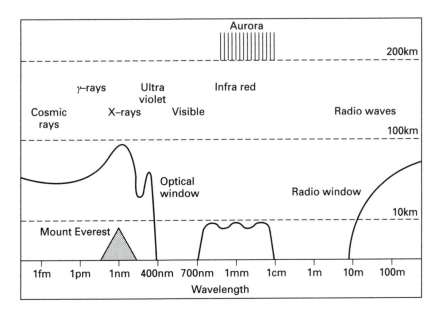

Fig. 7.1. Relative penetration of different types of solar radiation into the atmosphere. Only certain radio frequencies and radiation around that of visible light penetrate as far as sea level.

Table 7.1   Types of radiation in or near the visible window of the Earth's atmosphere

Wavelength	Type	Remarks
> 13 μm	Long-wave infrared	Between 13 and 17 μm, molecules like CO_2 and CFCs in the atmosphere absorb infrared (IR) radiation reflected back from the Earth to provide the blanket or greenhouse effect of global warming.
700 nm–13 μm	Short-wave infrared [a]	13 μm is the upper limit for IR radiation to penetrate to the Earth's surface from the Sun. This IR band heats up surfaces on the planet and also permits reflection of IR of < 13 μm back out into space.
400–700 nm	Visible spectrum[b]	Blue (*ca.* 400 nm) and red (*ca.* 700 nm) components are most important for photosynthesis while vision is most sensitive in green-yellow (500 nm) light.
315–400 nm	UV-A[c]	Long-wave ultraviolet not normally associated with injury to biological systems. Certain photoreceptors of animals and plants operate in this region.
280–315 nm	UV-B[d]	The wavelengths of sunlight likely to cause biological effects. There is some dispute over the upper limit: 315 nm is in the UNEP Report (1991) – others quote 320 nm.
< 280 nm	UV-C	These wavelengths are completely absorbed by the atmosphere before sunlight reaches the surface of the Earth. They are more disruptive to biological systems than UV-B and are commonly used for sterilization procedures.

[a–d] In relative terms these are 42.5, 51.3, 5.7 and 0.5% of total flux through the optical window, respectively.

may. Most incoming infrared radiation is completely absorbed before 10 km up (i.e. just above the height of Mount Everest) but small amounts of short wavelength infrared radiation, all the visible spectrum, and longer wavelength ultraviolet (UV) radiation reach sea level. Shorter wavelengths than these (X-rays, γ-rays and most cosmic rays), however, are screened out between 10 and 100 km above (Fig. 7.1).

Conventionally, UV light is subdivided into three categories (Table 7.1) and it is only UV-A and sometimes UV-B that penetrate as far as the surface of the Earth. Fortunately, the biologically more harmful UV-C is filtered out completely.

A variety of factors influence the atmospheric penetration of UV-B – the solar radiation known to cause most of the adverse effects upon biological systems. The major atmospheric barriers to UV-B penetration are the O_3 molecules in the stratosphere but other environmental factors also affect UV-B fluxes. These include tropospheric pollutants (other than O_3), sunspot activity on the Sun (which may increase stratospheric O_3 levels by nearly 2%), and reflections of UV-B from surfaces, clouds and aerosols. All are highly variable and account for the fact that long-term records of UV-B fluxes vary and often conflict. These large variations have prevented the establishment of a reliable global model to predict trends in UV-B flux. Indeed, networks of UV-B monitors, usually placed in urban areas for convenience, often record decreases of UV-B due to increases in tropospheric pollutants and cloudiness. The only consistent trends of enhanced fluxes of UV-B due to stratospheric O_3 depletion have been recorded by instruments in polar regions (see later).

Measuring techniques for UV-B vary as do the units of radiation flux. One common method of measuring relative change of atmospheric O_3 in the air column from ground level upwards is expressed as a notional sea-level thickness of gaseous layers consisting entirely of O_3. According to this method, a Dobson unit is equal to 0.0001 m at standard atmospheric temperature and pressure. This means a typical reading of 360 Dobson units is equivalent to a notional layer of O_3 which is 3.6 mm thick. Deviations from values like this are greatest over the South Pole in spring (September–November). First detected in 1982 by a ground survey team led by Dr Farman of the British Antarctic Survey, they were later substantiated in 1987 by satellites operated by NASA. Since then, progressively lower thicknesses of O_3 and commensurate increases in UV-B penetration have been recorded at this sensitive time. In 1992, for example, numbers of recorded Dobson units fell to 170 (i.e. *ca.* 50% of normal).

As a rough approximation, for every 5% stratospheric O_3 depletion there is a 10% increase in UV-B flux at sea level. Figure 7.2 shows the possible changes in UV-B fluxes per decade over different parts of the planet. Future increases of UV-B are likely to be greatest over the South Pole but a 7% increase per decade over most of the heavily populated regions of the northern hemisphere (35–60°N) is confidently predicted.

Monatomic oxygen, ozone, and hydroxyl radicals

In the thermosphere, 80 km and upwards above the surface of the Earth, oxygen exists almost exclusively in monatomic form because the high

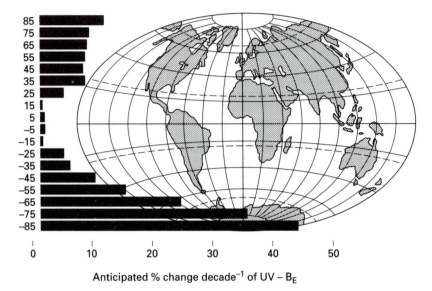

Fig. 7.2. Projected percentage increases in UV-B light fluxes per decade at different parts of the Earth's surface (adapted from data in the 1991 UNEP Report).

energy photons of sunlight at wavelengths less than 242 nm break up the most stable of molecules (Reaction 7.1; see also Fig. 7.3). In the stratosphere below, some of these oxygen atoms then combine with molecular oxygen to form O_3 (Reaction 7.2). Sunlight also causes some destruction of this O_3 layer (Reaction 7.3) but this particular reaction is rather slow. Consequently, greatest concentrations of O_3 are to be found in the stratosphere (15–40 km high; see Figs 1.2 and 1.3).

$$O_2 + light\ (<242\ nm) \Rightarrow 2O \tag{7.1}$$

$$O + O_2 + M^1 \Rightarrow O_3 + M \tag{7.2}$$

$$O_3 + light \Rightarrow O + O_2 \tag{7.3}$$

The most reactive free radicals in the atmosphere are $^{\bullet}OH$ radicals which are formed by a variety of reactions, including that of atomic oxygen with water vapour (Reaction 7.4). These $^{\bullet}OH$ radicals react with many gases in the atmosphere, including O_3 (Reaction 7.5), which means that a natural balance between O_3, water vapour, and free radicals like $HO_2^{\bullet}$, $^{\bullet}OH$, etc., exists.

[1] See Reactions 2.2 and 2.7.

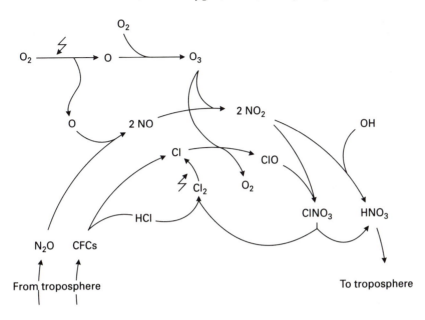

Fig. 7.3. Incoming solar radiation (< 242 nm) affects a number of upper atmospheric reactions (Z-shaped arrows) including the formation of atomic oxygen which gives rise to the stratospheric O_3 layer. Active O also breaks up N_2O, which arises from soil denitrification (see Chapter 3), to form NO which consumes O_3 as it is converted to NO_2. Strong sunlight also breaks up CFCs to form chlorine (and bromine) atoms which also consume O_3 as they are converted to chlorine (and bromine) oxides and nitrate. The chlorine and bromine are then recycled by additional light-dependent reactions so that they continue to remove O_3 molecules. N.B. It is important to appreciate that loss or uptake of monatomic oxygen (O) is the equivalent to O_3 removal because they are readily interconverted.

$$O + H_2O \Rightarrow 2 \cdot OH \qquad (7.4)$$

$$O_3 + {}^\cdot OH \Rightarrow HO_2{}^\cdot + O_2 \qquad (7.5)$$

Atmospheric nitrous oxide (N_2O) arises mainly from soil denitrification, especially if unused artificial fertilizers persist (Chapter 3), and, being inert, usually rises through the troposphere to the stratosphere without reacting on the way. Reaction with atomic oxygen, however, is capable of breaking open N_2O molecules (Fig. 7.3) to release NO (Reaction 7.6) which disturbs the natural equilibrium between O_3, water vapour, and $\cdot OH$ radicals. This accelerates the disappearance of O_3, either directly (Reactions 7.7 and 7.8), or by forming additional $\cdot OH$

radicals and NO_2 by reaction of $HO_2^{\bullet}$ with NO (Reaction 7.9). The NO_2 is then finally converted to nitric acid (HNO_3; Reaction 7.10).

$$N_2O + O \Rightarrow 2NO \tag{7.6}$$

$$NO + O_3 \Rightarrow NO_2 + O_2 \tag{7.7}$$

$$NO_2 + O \Rightarrow NO + O_2 \tag{7.8}$$

$$NO + HO_2^{\bullet} \Rightarrow NO_2 + {}^{\bullet}OH \tag{7.9}$$

$$NO_2 + {}^{\bullet}OH + M\,{}^{1} \Rightarrow HNO_3 + M \tag{7.10}$$

In the mid 1970s, considerable concern was expressed that large numbers of supersonic flights would inject significant quantities of NO and NO_2 into the lower stratosphere which would destroy O_3 and remove the protective shield to UV-B that this layer affords to the biosphere. For a number of reasons, technical as well as economic, the anticipated numbers of supersonic flights did not occur. Certain reassessments of potential hazards to the O_3 layer from supersonic and hypersonic aircraft in the stratosphere now indicate that they are still present but recovery would be relatively quick (i.e. 3 or so years). More of a problem from aircraft flying in the upper troposphere arises from unburnt hydrocarbons and nitrogen oxides forming additional O_3 which adds to global warming (see Chapter 8).

The main threats to the stratospheric O_3 layer, therefore, come from N_2O (see above) and halocarbons from aerosol cans, escapes of coolants, burning of insulation materials, etc. (see below). The recovery time in each case is from 20 to 50 years – too long for short-term remedial action.

Sprays, refrigerants, insulation materials, and solvents

The last two decades have seen a dramatic increase in the commercial production of halocarbons which contain only fluorine, chlorine, or bromine as well as carbon. Most contain only the two lighter elements and are known as chlorofluorocarbons (CFCs). CFCs are now used widely as aerosol propellants, refrigerator coolants, foam blowing agents for the production of insulation and packing materials, etc., solvents to clean electronic components or to thin and strip paints, or medically as anaesthetics or for spray inhalants for the treatment of asthmatics and angina pectoris attacks.

---

[1] See Reactions 2.2 and 2.7.

Originally thought to be completely inert and harmless, compounds such as $CFCl_3$ (CFC-11) and CF_2Cl_2 (CFC-12), as well as smaller amounts of CF_4, C_2F_6, CCl_4, $CClF_3$, $CHClF_2$, $CHCl_2F$, $C_3H_3Cl_3$, $C_2Cl_3F_3$, $C_2Cl_4F_4$, C_2ClF_5, $CBrF_3$ and C_2BrClF_4 (see also Table 7.2), measured as chlorine, now amount to just over 4 nl l^{-1} of the total atmosphere – a seemingly small amount but which actually accounts for almost the whole of the world's production of CFCs since they were first introduced. This is due entirely to their inertness which was originally thought to be their greatest advantage.

Estimated atmospheric lifetimes of these compounds are very long. CFC-115, for example, has an estimated lifetime of 548 years. Once degraded to chlorine atoms, etc., their lifetime is still quite long (1–2 years) and they go on repeatedly destroying O_3. One chlorine atom in the stratosphere can destroy several thousand O_3 molecules and a bromine atom destroys 40 times as many O_3 molecules as a chlorine atom does.

Problems with CFCs and halons start in the stratosphere. By a complicated series of reactions, they are broken down by light or by reaction with monatomic oxygen to produce chlorine and bromine atoms. These react with (and remove) O_3 (and monatomic oxygen) from the stratosphere (Reactions 7.11 and 7.12) by forming chlorine and bromine oxides (ClO and BrO) in mechanisms very similar to those involving the nitrogen oxides (Reactions 7.7 and 7.8) as shown in Fig. 7.3. In the final stages, the nitrogen oxides react with chlorine and bromine oxides to produce highly unusual chlorine (and bromine) nitrate (e.g. $ClNO_3$; Reaction 7.13). These react with water to form nitric acid (Reactions 7.14–15) which locks away nitrogen oxides as a mechanism of scavenging chlorine and bromine atoms if it becomes frozen (see below).

$$Cl + O_3 \Rightarrow ClO + O_2 \qquad (7.11)$$

$$ClO + O \Rightarrow Cl + O_2 \qquad (7.12)$$

$$ClO + NO_2 + M^1 \Rightarrow ClNO_3 + M \qquad (7.13)$$

$$2ClNO_3 + 2H_2O \Rightarrow 2HNO_3 + 2HCl + O_2 \qquad (7.14)$$

Polar vortices

Rates of stratospheric O_3 depletion increase dramatically in the spring of each year over Antarctica (September–November). The appearance of

[1] See Reactions 2.2 and 2.7.

Table 7.2 Formulae and common abbreviations of halons, CFCs and HCFCs mentioned in the 1990 London Amendment to the 1987 Montreal Agreement plus some of the possible HFA replacements

Annex[a]	Group	Substance	Common code	O_3 depleting potential[b] S. state	10 years
A	I	$CFCl_3$	CFC-11	1.0	1.0
		CF_2Cl_2	CFC-12	0.8	
		$CF_2ClCFCl_2$	CFC-113	1.1	1.25
		CF_2ClCF_2Cl	CFC-114	0.8	
		CF_3CF_2Cl	CFC-115	0.4	
	II	CF_2ClBr	Halon-1211	4.1	10.5
		CF_3Br	Halon-1301	12.5	10.4
		CF_3CFBr_2	Halon-2402	5.9	12.2
B	I	CF_3Cl	CFC-13	ca. 1	
		$CFCl_2CCl_3$	CFC-111	ca. 1	
		$CFCl_2CFCl_2$	CFC-112	ca. 1	
		Series from C_3FCl_7 to C_3F_7Cl	Codes from CFC-211 to CFC-217	ca. 1	
	II	CCl_4	Carbon tetrachloride	1.08	1.25
	III	CH_3CCl_3	Methyl chloroform	0.12	0.75
C	I	$CHFCl_2$	HCFC-21	< 0.2	
		CHF_2Cl	HCFC-22	0.05	0.17
		CH_2FCl	HCFC-31	< 0.2	
		Various compounds from C_2HFCl_4 to C_2H_4FCl	Codes from HCFC-121 to HCFC-151	< 0.2	
		from C_3HFCl_6 to C_3H_6FCl	HCFC-221 to HCFC-271	< 0.2	
Unclassified		CH_3Br	Methyl bromide	0.57	5.4
(Possible replacements)		C_4H_8s	Butanes	0	
		CHF_2CF_3	HFA-125	0	
		CHF_2CHF_2	HFA-134	0	
		CH_2FCF_3	HFA-134a	0	
		CH_3CHF_2	HFA-152c	0	

[a] Annexes A and B consist of controlled compounds for the purposes of the Montreal Protocol, and Annex C lists substances to be regarded as transitional and to be monitored closely as HFA replacements become available.
[b] O_3 depleting potential calculated with that for CFC-11 set to 1.0 for the achievement of steady state levels (S. state) and within 10 years. With the very long lifetimes of CFCs, etc., in the literature, the latter (where available) are more realistic. Approximate steady state values are either those from the UNEP (1991) report or, where available, more accurate values taken from Jones & Wigley (1989), p.20, and Solomon & Albritton, 1992. The 10-year values are those of Solomon & Albritton, 1992.

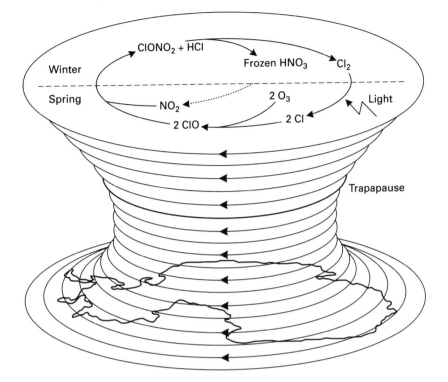

Fig. 7.4. Diagram to illustrate the isolation of the air mass inside the Antarctic vortex from that of the rest of the southern hemisphere through the Antarctic winter. This vortex of high-speed winds extends above the tropopause and only breaks up when light begins to penetrate in spring. Temperatures inside the vortex are very low – sufficient to freeze out HNO_3, with the result that ˙OH radicals and chlorine atoms accumulate overwinter. Once the vortex breaks up these reactants attack O_3 atoms and create a significant O_3-hole over Antarctica which now often extends over the southern tip of South America.

this O_3 hole (Plate 9) coincides with the breakup of the vortex of rapid air movements (Fig. 7.4) which extends from ground level into the stratosphere and isolates the whole atmosphere over Antarctica from the rest of the air movements in the southern hemisphere through the polar winter (June–September). Inside this vortex of high speed winds, little or no sunlight penetrates and temperatures can fall below $-60°C$. Normally, nitric acid reacts with ˙OH radicals to produce NO_2 (Reaction 7.16) but, at these very low temperatures, nitric acid freezes into solid aerosols, which effectively removes both the chlorine and bromine scavenging effect of the nitrogen oxides (see above) and this ˙OH depleting reaction. In spring, as the vortex breaks up and sunlight begins to penetrate again, the extra chlorine and bromine atoms and ˙OH radicals then rapidly destroy stratospheric O_3 (Reaction 7.5; Fig. 7.4).

$$ClNO_3 + HCl \Rightarrow Cl_2 + HNO_3 \qquad\qquad (7.15)$$

$$HNO_3 + 2˙OH \Rightarrow NO_2 + H_2O + O_2 \qquad\qquad (7.16)$$

Over the Arctic pole, winter temperatures do not fall as low as those over Antarctica nor does any vortex that forms extend into the stratosphere and isolate the entrapped air mass so completely. As a result, nitric acid does not freeze out so readily and high air movements overwinter may still bring in compounds which remove ˙OH radicals and scavenge chlorine and bromine atoms. This means that springtime O_3 depletion over the Arctic, although significant, is never likely to be as severe as over Antarctica.

Alternatives to CFCs

The first CFCs, $CFCl_3$ (CFC-11) and CF_2Cl_2 (CFC-12), were synthesized by chemists of General Motors as possible replacement coolants as long ago as 1928. Today, overall production of these two CFCs still dominates the CFC market – closely followed by significant quantities of $CF_2ClCFCl_2$ (CFC-113), CF_2ClCF_2Cl (CFC-114) and CF_3CF_2Cl (CFC-115). All this, according to the 1987 Montreal Protocol as amended in London in 1990 and Copenhagen in 1992, is about to change. This international treaty, now signed by over 50 nations, endeavours to cap CFC production at 1986 levels and, thereafter, progressively reduce their synthesis – first by a 50% cut in production by January 1995 followed by an 85% reduction by 1997 and complete phase-out by year 2000. In addition, substances in Annexes A (Group I) and B (II and III) of the treaty (i.e. the five above plus the halons and the solvents CCl_4 and CH_3CCl_3 – see Table 7.2) are subject to both export and import bans. Similar, but later, controls on CFCs in Annex B (Group I) are also to be imposed. A third group (Annex C) of hydrochlorofluorocarbons (HCFCs) is to be closely monitored and assessed with the word 'transitional' in mind as chlorine-free alternatives are sought and introduced – phase-out of HCFCs not being anticipated before 2020. The situation with methyl bromide (CH_3Br), a pesticide, is still unsatisfactory. It was not mentioned in the Montreal agreement nor the London amendment but the 1992 Copenhagen meeting agreed to freeze production at 1991 levels by 1995 without setting a final date for phasing-out CH_3Br. This still appears to be inadequate given the fact that bromine is more than 40 times more effective at removing stratospheric O_3 than chlorine.

The stratospheric O_3-destroying potential of HCFCs is roughly one-fifth that of CFCs because their reactivity in the troposphere is much greater and they tend to release their chlorine in the troposphere rather than the stratosphere. Indeed, to be on the safe side, many chemical companies have sought compounds containing no chlorine at all. The best alternatives in this direction appear to be hydrogen fluoroalkanes (HFAs) such as CHF_2CHF_2 (HFA-134) and CH_2FCF_3 (HFA-134a). Unfortunately, there has also been a confusing trend, intentional or otherwise, to relabel certain HCFCs as HFAs despite their being clearly designated as chlorine-containing HCFCs in Annex C of the Montreal Protocol (see Table 7.2) and subject to the close monitoring specified there. These include CHF_2Cl (HCFC-22), $CHCl_2CF_3$ (HCFC-123) and CH_3CFCl_2 (HCFC-141b).

Finding suitable replacements for CFCs is difficult. HCFCs and HFAs are less stable than CFCs and, generally, more flammable. Moreover, they do not have the advantage of always being non-toxic nor do they have the very low thermal conductivities (i.e. high insulation values) of CFCs. It is the introduction of hydrogen atoms into CFCs that introduces instability. Consequently, HCFCs react more readily with the ˙OH radicals of the troposphere to form H_2O and chlorine-containing radicals which then form HCl and HF rather than O_3-destroying chlorine atoms. HFAs, by contrast, only form small quantities of HF.

Potential problems do not end with the elimination of chlorine (and bromine). Butanes (C_4H_8), for example, are often used as 'O$_3$-friendly' aerosol propellants but they are also highly flammable and have already caused several tragic accidents. HFAs with low fluorine contents are little better.

Even HFAs with higher fluorine contents have potential problems. The asymmetric HFA-134a (CH_2FCF_3), like certain other HCFCs (e.g. HCFC-123 or CF_3CHCl_2), produces trifluoroacetate (CF_3COO^-) as well as HF upon reaction with ˙OH radicals in the troposphere. Currently, little is known of the fate of CF_3COO^- in the atmosphere or as it is rained out onto the biosphere. Fears have been expressed that it could be defluorinated to CH_2FCOO^- which is known to be highly toxic to humans. Such a pathway is less likely from the symmetrical alternative HFA-134 (CHF_2CHF_2) which only decays to CO_2 and HF in the lower atmosphere.

The limitations specified in the 1987 Montreal protocol as amended in London in 1990 will still permit levels of O_3-destroying chlorine atoms to rise in the stratosphere (Scenario B, Fig. 7.5). International agreement is still required after the Copenhagen meeting in 1992 to enforce further phasing out of CFC production. Figure 7.5 projects the effect of this (Scenario C), assuming that true HFAs and less harmful HCFCs replace CFCs entirely, that natural emissions (mainly CH_3Cl) remain constant,

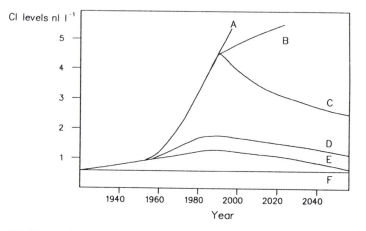

Fig. 7.5. Past and projected global levels of chlorine atoms, assuming that natural emissions (mainly CH_3Cl from seaweeds) remain constant. Scenario **A** indicates what the trend would have been without the Montreal Protocol and Scenario **B** predicts what is possible if the recommendations of the Copenhagen amendment of 1992 are effective. If all CFCs, apart from very limited medical uses, are actually phased out and replaced by HFAs, then Scenario **C** may be possible but this assumes that the toxic solvent CCl_4 is phased out completely and the use of the alternative solvent CH_3CCl_3 does not increase. N.B. Chlorine levels due to natural releases (mainly CH_3Cl) are described by **F**, those of CCl_4 and CH_3CCl_3 over F by **E** and **D** respectively.

that the toxic solvent CCl_4 is completely phased out, and that demand for the less toxic alternative CH_3CCl_3 remains constant.

Clearly, urgent research is also required to identify the relative merits or demerits of alternative HCFCs and HFAs so that fresh problems do not supplant the earlier mistakes.

Biological action spectra

A biological action spectrum consists of a series of measurements of a particular biological function (e.g. rate of photosynthesis, skin burn or erythema, etc.) across a range of wavelengths. In the case of UV-B studies, this may involve wavelengths below 290 nm way up into the visible spectrum. Such action spectra are crucial for an understanding of the consequences of stratospheric O_3 depletion on the biosphere and can be used in a variety of different ways. For example, they are used to calculate radiation amplification factors (RAFs) which, in the case of UV-B, are a measure of the biological significance of a solar irradiation

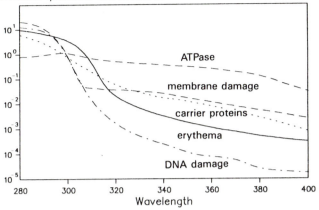

Relative quantum effectiveness

Fig. 7.6. Action spectra of some important biological processes ranging from DNA dimerization through sunburn (erythema) to effects on transport processes in, on and across membranes (e.g. ATPases).

change over a specific wavelength range (e.g. 290–315 nm) brought about by a particular amount of O_3 depletion. RAFs can also be used to determine gradients of biologically effective UV-B ($UV\text{-}B_{BE}$) across different latitudes or to compare sunlight with artificial light sources used in $UV\text{-}B_{BE}$ experimentation. They also allow relative assessments of different biological processes and permit comparisons between species.

Action spectra are affected both by external environmental factors and by the internal properties of the tissues penetrated. They are also invaluable in determining which biological compound or chromophore, if any, absorbs $UV\text{-}B_{BE}$. If this can be achieved then the absorption spectra of these chromophores must be determined both *in vitro* and *in vivo*. Rarely, however, is this a simple situation with a single chromophore – usually several overlap. In plants, for example, a wide range of flavonoids (complex phenols), as well as proteins and nucleic acids, absorb $UV\text{-}B_{BE}$.

As a general rule, it is important to identify those factors, components, or interactions that steepen any curve of $UV\text{-}B_{BE}$ response, rather than flatten it, when assessing the implications of O_3 depletion. This procedure also permits the early discrimination of important components of biological response. Figure 7.6, for example, indicates that DNA changes are more important than those causing sunburn (erythema) which, in turn, are more important than changes in membrane transport (ATPase activity). They can also provide more detailed information. For example, action spectra for inducing experimental cancers in shaved or mutant mice with UV-B light are almost identical to those shown for DNA in Fig. 7.6. The implication of this is that the early mechanisms responsible for starting skin cancers involve changes in DNA.

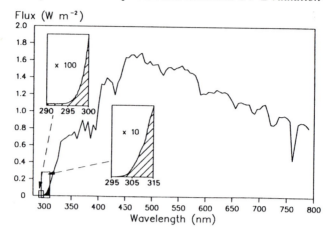

Fig. 7.7. Total flux of UV and visible components of sunlight on a clear day in midsummer (courtesy of Dr Nigel Paul, Lancaster). Note the very small proportion of UV-B relative to the rest, especially below 300 nm.

UV-B fluxes and experimental irradiation

Early experimentation often overestimated the biological effects of UV-B light. Unsuitable artificial light sources were used – often continuously or in a simple on–off 'square wave' mode and, in plant studies, without due regard to the total radiant flux from 290 to > 700 nm. Fluxes of UV-B$_{BE}$ vary through the day, as do total radiant fluxes, but not in strict register with each other. Environmental factors, such as hazes, snow cover and a variety of other factors, also affect the ratio of UV-B to UV-A plus visible radiation. Moreover, the relative changes in UV-B are very small by comparison to the total flux (Fig. 7.7) and vary not just through the day but according to season (Fig. 7.8).

This means experimental exposures have to simulate these changes as closely as possible if a 'claimed' biological response can be regarded as real. Technically this is very difficult and requires modulation of output of all the lamps involved. In modern field exposure systems, ambient fluxes of UV-B are now monitored and UV-B supplementation is added (to stimulate a particular O_3 depletion rate above this) as a constant ratio over the ambient level. In very advanced systems, closed loop computer control systems also correct for any shading the lights and their supports may have upon the species under test.

The reverse approach involves the use of filters instead of additional lights. Test species, often from higher latitudes and experimentally cooled, are exposed to ambient light and have filters placed above them which are translucent to UV-A and visible irradiation but absorb part of

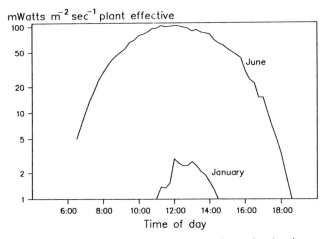

Fig. 7.8. Representative fluxes of UV-B$_{BE}$ throughout the day in summer and winter (courtesy of Dr Nigel Paul, Lancaster). Note that the scale of measurement has to be converted to a logarithmic one because the maximum flux at midday in midsummer is 50 times that at the same time in winter.

the UV-B component. With a range of filters, this allows the simulation of solar UV-B fluxes over a range of latitudes and altitudes. Such an approach has the advantage that other temperatures can be superimposed or additional air pollutants or elevated levels of CO_2 added.

For small animals, plants, microbes and mixed ecosystems, a third approach is often favoured. This usually involves a xenon light source and a range of filters which permits exposure of small areas (< 100 cm^2) to narrow 'slices' of UV radiation. Various contributions of other light can also be given. Such equipment is often used to establish biological action spectra, but subsequent interpretation can be difficult when translated to likely responses to UV-B$_{BE}$ in a continuous spectrum.

Higher plants

Well over 300 different plant species have been tested for responses to elevated levels of UV-B and changes have been detected in just over half of them. As a general rule, broad-leaved species are more vulnerable than narrow-leaved plants – cucurbits and brassicas are especially sensitive. With improvements to experimental techniques (see above), some reappraisal is now in progress but several biochemical, physiological and morphological responses have been confirmed (see Table 7.3). As might be expected, the early choice of species for study was biased towards agricultural crops, and studies on native trees or shrubs, and

Table 7.3 Summary of reported changes in plants due to enhanced UV-B fluxes (updated from Krupa & Kickert, 1989)

General type	Parameter (with comments)
Visible changes	Cuticular reflectivity (glazing) sometimes increased, leaf bronzing or reddening (rare), chlorosis (very rare)
Morphological	Leaf area (usually reduced), thickening, margins curling, leaf number or tillering in grasses (usually increased) stem elongation or stunting (varies between species), flower (and, hence, fruit) number (sometimes decreased), seedling heights affected (some stunting)
Physiological	Net photosynthesis (sometimes affected), dark respiration (occasionally increased)
Yield	Specific leaf weight, dry matter production, carbon allocation, crop yield, crop quality (usually all reduced)
Temporal	Bud burst and flowering sometimes advanced (rarely delayed), senescence sometimes delayed
Interactions with:	Light (enhanced photorepair), water availability, nutrients, heat and cold stresses, other air pollutants, enhanced CO_2, infection with fungi, predation by insects (all possible but not yet fully evaluated – most are additive rather than true interactions)
Competition	Between and within species (considerable variation from highly sensitive to tolerant – see Table 7.4)

possibly sensitive ecosystems such as heathlands or tundra, are still in progress.

With certain exceptions, symptoms of visible injury specifically due to UV-B in the field are rare. Often, they can be confused with those due to other stresses. However, experimental treatment of plants with enhanced levels of UV-B, typifying higher rates of stratospheric O_3 depletion than now, have shown quite distinctive responses. Some leaves thicken so the path length of light flux is greater and some increase their surface cuticular reflectivity (see Plate 13). Others form additional protective pigments, especially in the surface (epidermal) layers. Generally, these screening pigments are flavonoids (complex phenols) which frequently impart a bronzing or reddening to leaves.

The reflective UV-B-treated tobacco leaves shown in Plate 13 also show that the leaf margins curl upwards. This is probably due to enhanced UV-B destruction of a plant hormone (IAA) in the uppermost layers which causes the lower layers to grow more rapidly, forcing the leaf edges upwards.

Other morphological changes occur which do not imply reductions in productivity caused by increased reflection or absorption of light energy. Increased branching, shorter leaf lengths, and increased numbers of leaves due to elevated UV-B have been reported, especially for grasses, while bud-burst may occur earlier in spring and leaves stay longer on

Table 7.4 Summary of species sensitive and tolerant to UV-B on the basis of biomass accumulation (summarized, adapted, and updated from Krupa & Kickert, 1989)

Type	Sensitive	Intermediate	Tolerant
Fibre			Cotton
C$_3$ grain	Barley, oats	Rice, rye	Wheat, sunflower
C$_4$ grain		Sorghum	Maize, millet
Legumes	Soya, peas	Beans, peas	
Fruit	Tomato, cucumber, pumpkin, melon, raspberry	Pepper	Oranges, aubergines
Stem crops	Rhubarb, sugarcane		Celery, asparagus
Root crops	Sugarbeet, carrots	Potato	Radish, onions, parsnips
Other vegetables	Cauliflower, broccoli, chards, sprouts, mustard, spinach	Lettuce	Artichoke, cabbage, kohlrabi
Forage			Red clover, alfalfa
Flowers	Bluebell, *Coleus*	*Petunia*	Marigolds, *Poinsettia*
Weeds	Groundsel, plantains		Docks, wild oats
Trees	Birches	Oaks, beeches	Most conifers

plants in the autumn. Sometimes, seasonal changes are also evident in annuals and biennials. Flowering may be advanced or retarded, depending on species, but reduced numbers of flowers (and hence fruits) often occur.

Measurements of net photosynthesis vary widely in the presence and absence of additional UV-B radiation. These range from no change to 40% depressions, but they are often transient and only occur at certain stages of leaf development or after winter dormancy. Levels of pigments, rates of transpiration, and stomatal conductances are also affected. Detrimental effects of UV-B on growth have been demonstrated for certain species (Table 7.4) but others are unaffected. Many important crops, for example rice, show strong depressions of growth while many other cereals are less affected. Peas grown in the UK, for example, will probably show a 7% depression in final seed yield for a 7% expected increase in UV-B$_{BE}$ over the next decade (UNEP Report, 1991) for latitudes around 52°N. Variation in cultivar response within species, however, is very great. This means that, for most agricultural crops at least, plant breeding and selection of UV-B tolerant cultivars in the meantime can overcome potential losses. For long-lived species or for native plants in certain ecosystems, where the gene pool is restricted, differential changes in productivity between species, however, could be significant. Some species will be out-competed and replaced by those less affected by UV-B.

External environmental factors also affect plant response to enhanced UV-B light. For example, certain plants (e.g. maize at 32°C) growing at slightly higher temperatures do not show as much growth loss with extra UV-B as those growing at lower temperatures (e.g. maize at 28°C). This means, in a slightly warmer world, the effects of UV-B on plants may be less significant. Conversely, the extra growth of rice and wheat associated with 650 μl l^{-1} CO_2 rather than 350 μl l^{-1} disappears in the additional presence of UV-B but that of soya is unaffected. There is also evidence to suggest that plant response to heavy metals changes adversely in the presence of additional UV-B radiation. Similarly, the severity of certain plant diseases is influenced by additional UV-B radiation. To a certain extent, this depends on differential effects of UV-B on both host and pathogen as well as the timing of the UV-B exposure and infection.

The mechanisms by which UV-B affects plants are now under close study. Plant response in terms of protection is known to include induction of enzymes associated with the fresh synthesis of UV-B protective flavonoids in epidermal layers or enhanced repair of DNA dimers by a process known as photoreactivation. Apart from DNA, which is known to be highly sensitive in most biological systems to UV-B$_{BE}$, the integrity of cell membranes is also affected. Enhanced cell leakiness and the appearance of malondialdehyde (see Chapter 6) are coupled to the formation of extra antioxidants, and other indicators of environmental stress (e.g. polyamines) are indicative of enhanced free radical and HHP formation associated with general oxidative stress. Taking into account the lack of specific photosynthetic effects to account for measurable losses in growth, the energy-diversion hypothesis (see earlier chapters) is the most likely alternative explanation. Changes of nucleic acids, membranes, and UV-B absorbing hormones, etc., in cells exposed to additional UV-B means that additional energy must be expended to repair these defects and this redirected energy is therefore unavailable for growth.

Microbe effects

Most of the early understanding of the biological effects of UV has come from bacteriological studies of the sterilizing effects of UV-C. However, disturbances to nucleic acids are not confined to UV-C. Both UV-C (256 nm) and UV-B (313 nm) may induce adjacent pyrimidine bases in the same DNA strand to pair with each other rather than with their normal purine base partners in the opposite strand. If these pyrimidine dimers are not corrected by a process known as photoreactivation (also started by UV-C and UV-B radiation) then a faulty genetic message is carried either into the next generation when that organism reproduces or into adjacent cells when tissue cells divide. In animals, this sequence is often linked with cancer initiation.

UV-A and visible light also influence recovery from UV-B-induced injury to DNA by causing an additional growth delay which allows more time for the photoreactivation process to be induced and take place. However, if both UV-A and visible radiation are very strong, an extra sensitization reaction may also interfere with photoreactive repair.

Some microbes, such as yeasts or cyanobacteria, have additional mechanisms of UV-B protection. In the presence of UV-B (303 nm), for example, they often release significant UV-B-absorbing phenols into the medium around them. Germinating spores (uredospores) from certain rust fungi (e.g. *Puccinia striiformis*) are very sensitive to UV-B while those from other species (e.g. *P. graminis*) are more resistant. Again, the sensitivity is linked to UV-B-induced changes in nucleic acids, especially when the humidity is high, but the resistance of *P. graminis* uredospores is probably due to additional UV-B-absorbing carotenoid pigments.

Chapters 3 and 4 have already emphasized the global importance of microbes. Nearly all biological N_2 fixation is carried out by microbes, mainly cyanobacteria, while in soils, inside higher plants, or in shallow waters (e.g. rice paddies). Moreover, many free-living cyanobacteria appear to be UV-B sensitive. More information on them is required urgently because, if stratospheric O_3 depletion continues, one of the important effects of UV-B over land and in estuarine waters could be on global nitrogen cycling.

Marine life

Penetration of light into ocean waters depends on wavelength. Blue light (460 nm) penetrates as far down as 260 m but red light (700 nm) much less (< 2 m). Studies of UV-B penetration into columns of seawater are still in progress but it is now evident that UV-B may go down as far as 65 m in Antarctic waters.

The consequences of this for marine phytoplankton are considerable. This wide range of photosynthetic microbes dwell largely in the photic zone and orientate themselves by both light and gravity. This allows them to optimize light (mainly blue) interception and, thereby, maximize photosynthetic growth. In global terms, these amounts are vast – 104 Gt a^{-1} – slightly more than the whole of land-based photosynthetic productivity. Moreover, over 70% of this productivity is concentrated in polar waters where the rate of change of UV-B$_{BE}$ per decade is greatest (Fig. 7.2).

The ability of phytoplankton to orientate themselves in the water column is also affected by UV-B. This, in turn, exposes some of the phytoplankton species to more damaging levels of UV-B. However, little is known of the mechanisms of UV-B injury in phytoplankton but current evidence suggests that they differ from those in land-based ecosystems. The action spectrum of phytoplankton mobility over the UV-B plus UV-A regions differs from other action spectra, and the photoreceptors of phytoplankton are more sensitive than those in land-based plants because they have to trap light at low levels more efficiently. In shallow waters, some phytoplankton have protective pigments such as unusual carotenoids or mycosporine-like amino acids which absorb radiation in the UV-A and blue light region. Some of these adaptations are unhelpful in deeper oceanic waters where additional light gathering for photosynthesis requires this additional blue light. The result is that, while reductions in rates of photosynthesis at the surface due to UV-B might be in the range of 15–20%, those phytoplankton lower down (> 5 m) experience 40–60% inhibition. Consequently, fears that regular and deepening O_3 holes over Antarctica may reduce overall phytoplankton productivity and adversely effect the whole food chain in the area are still justified, although overall estimates have been scaled down. Certain measurements under the 1989 O_3 hole suggested depressions of up to 25% but studies in 1991 reduced this to 6–12%.

Chapter 4 has emphasized the importance of oceanic CH_3SCH_3 emissions for the nucleation of cloud formation. These CH_3SCH_3 emissions are directly related to the decay of marine phytoplankton. Any process like UV-B enhancement which influences phytoplankton productivity will affect CH_3SCH_3 releases, and, hence, the rate of cloud formation over the oceans. The consequences for this in terms of global warming are discussed in the next chapter.

Some, but not all, zooplankton are also sensitive to UV-B radiation. Some are well protected by a range of pigments and others have an increased capacity for DNA repair. This means that there are wide differences in estimates of death rates between similar species. For a 15% stratospheric O_3 depletion, 50% of the larvae of some shrimp species may die but those of other shrimp species would be unaffected. Often these differences are due to reproduction behaviour. Those species most likely to be affected have a seasonal cycle which coincides with the highest radiant fluxes into oceanic surface waters.

A similar situation exists for fish fry. Anchovy larvae at a depth of 0.5 m would be killed by UV-B fluxes equivalent to 10% O_3 depletion. The majority of mature fish, on the other hand, are likely to be affected only indirectly by a decline in marine phytoplankton. However, a form of erythema has been detected in the upward-facing skin area of farmed fish kept in large enclosures at depths much less than they would normally swim in the wild.

Human health

UV-B light is beneficial to humans in the sense that it converts 7-dehydro-cholesterol in skin cells into vitamin D_3 which supplements the dietary intake of vitamin D_2 from the diet – both of which are required for correct bone formation. UV-B radiation also suppresses certain allergic responses of the skin. For example, some infantile eczema rashes on the insides of elbows or knees improve for a time when exposed to more sunlight. However, most other effects of UV-B on humans are either directly or indirectly harmful. The list of medical conditions known to be aggravated by UV-B continues to grow (Table 7.5).

Skin cancers

Incidence rates of all the major types of skin cancer are related to exposure to solar radiation. Indeed, skin cancers are the commonest forms of all cancers in fair-skinned populations (over 40%) and their frequency, as a proportion of the whole, continues to grow. This trend is linked to behavioural changes associated with increased affluence whereby large numbers of fair-skinned individuals from higher latitudes travel to lower latitudes on holiday and expose more of their skin for longer periods to higher light intensities.

Skin cancers are of two basic types – melanomas and non-melanomatous carcinomas. The former develop in the pigment-forming cells of the skin, the latter in keratin-forming cells. The incidence of melanomas is rising, especially in fair-skinned individuals over 40, and is positively linked both with the degree of freckling of an individual and with intermittent, rather than continuous, exposures to high levels of sunlight from an early age onwards. Melanomas are especially prevalent in immigrant groups who have moved from high to low latitudes (e.g. the Irish in Australia) and in those who suffer the disease xeroderma pigmentosum who are genetically deficient in photoactivated DNA repair.

Melanomas often cause death. Survival rates from melanomas differ between men (45%) and women (66%). To a certain extent this is compensated by a lower incidence rate for males. Mortality rates for other forms of skin cancers also vary but are much less than for melanomas. In areas where modern medical treatment is available, they are usually less than 1%. Basal cell carcinomas are the main type of non-melanomatous skin cancer and four times more common than squamous cell carcinomas. The latter, however, cause more deaths, especially in fair-skinned individuals.

Calculation of cancer risks due to enhanced UV-B exposure is extremely difficult. Both the radiation amplification factors (RAFs, see

Table 7.5 Medical conditions known to be induced by enhanced UV-B and their likelihood of increase with further stratospheric O_3 depletion

Area affected	Condition or agent	Individual types	RAF[a]
Skin	Erythema (sunburn)		1.7
	Non-melanomatous cancer	Basal cell carcinoma	
		Squamous cell carcinoma	1.4
		Lip cancer	
		Salivary gland cancer	
	Melanoma	Superficial spreading	
	(Cutaneous malignant)	Lentigo maligna	
		(Hutchinson's freckle)	1.6
		Nodular	
		Unclassified	
	Xeroderma pigmentosa	(Conditioned worsened)	
	Skin ageing	Elastosis	1.2
Eye	Corneal damage	Acute photokeratitis	1.1
		(snow-blindness)	
	Lens abnormalities	Nuclear cataracts	
		Posterior subcapsular	0.7
		cataracts	
		Cortical cataracts	
		Presbyopia (reading	
		glasses required earlier)	
	Other effects	Intraocular melanoma	
Infections[b]	Viruses	Measles	
		Chicken pox	
		Herpes simplex	
		HIV-1 activation	
	Protozoa	Leishmaniasis	0.8
		Malaria	
	Bacteria	Tuberculosis	
		Leprosy	
	Fungi	Candidiasis (thrush)	

[a] Latest Radiation Amplification Factors given in 1991 UNEP report.
[b] Those where there is a likelihood of altered immunosuppressive response because they have a stage which involves the skin (e.g. rashes).

earlier) for photocarcinogenesis and the percentage increase in skin cancer that results from each increment in annual UV dose (biological amplification factors; BAFs) for the different types of cancers have to be taken into account. The most recent UNEP estimates claim that a 10% increase in stratospheric O_3 depletion is likely to cause a 26% increase in numbers of non-melanoma skin cases (300 000 in total) and 4500 extra melanoma cases worldwide. These are salutary numbers but significantly lower than earlier estimates.

Eye damage

The cornea, lens, and inner parts of the eye are all adversely affected by UV-B radiation (see Table 7.5). Cancers inside the eye are rare but fluxes of solar radiation and the incidence rates of intraocular melanoma are interrelated. Blue-eyed individuals appear to be most susceptible, especially if they live at low latitudes.

Reddening of the front of the eye, the eyelids, and the skin around the eye (photokeratitis) is commonly called snow-blindness. Peak sensitivity to snow-blindness is around 307 nm and protective glasses should be worn at all times at high altitudes, especially in snow, to avoid damage not just to the outside of the cornea but to the inner layers of the cornea as well. Unlike the tanning of skin, no tolerance is ever acquired. In fact, the reverse is the case – repeated exposures increase sensitivity because the repair processes slow down.

Cataracts are diseases of the lens which depend on both latitude and age. At least three different types of cataracts are known (see Table 7.5), all of which are aggravated by UV-B. If detected early they can be treated but, unfortunately, they remain the greatest single case of avoidable blindness worldwide and account for over 50% of the world's 33m blind. The mechanisms by which UV-B disturbs the translucency of the lens are still not understood, but the major protein of the lens, β-crystallin, is normally tightly stacked rather like planks in a timber-yard. UV-B radiation causes these stacks to become disorganized and this causes the lens to cloud over.

Predictions for cataracts worldwide due to enhanced UV-B indicate an increase, especially for the old. The EPA calculated in 1985 that those US citizens born in the years 1986–2029 will develop 4.3m extra cataracts if stratospheric O_3 is depleted by 10%. Most of these will be detected and corrected but many of those affected in developing countries will not be so fortunate.

Infections and immune responses

UV-B radiation adversely affects the ability of the skin's immune system to respond adequately to foreign bodies such as bacteria, viruses, fungi, and foreign chemicals (xenobiotics). Normally, many of these act as antigens which elicit the production of antibodies as well as an immune response between antigen-exposed cells and certain lymphocytes known as effector T-cells. Exposure to UV light, however, causes antigen-exposed cells to induce T suppressor lymphocytes (T_s cells) to appear instead. Normally, T_s cells are produced by the body so it can distinguish itself from an invader. This means that, when potential antigens arrive in the presence of extra UV-B, the body is unable to recognize them as foreign and the full immune suppression response does not take place. Consequently, the incidence rates and severities of infection are much worse.

A wide range of bacterial, viral and fungal infections which involve the skin at some stage during their progress are thought to be influenced by additional UV-B radiation (see Table 7.5). Many of these are of worldwide significance. Moreover, vaccination programmes often rely upon immunization. This means that an individual exposed to extra UV-B may not acquire the necessary immunity required when injected with a highly weakened antigen.

Animal effects

Most mammals have dense fur to protect them from UV-B radiation, but some non-melanomatous skin cancers do occur in animals with less hair or, experimentally, with shaved or hairless mutant animals exposed to UV-B. They do, however, suffer eye damage like humans. For example, enhanced levels of UV-B increase the severity of eye infections and the numbers of eye cancers in cattle. In the Magallanes, which is the southernmost region of Chile, high rates of temporary blindness in sheep, rabbits and horses, with symptoms rather like snow-blindness, have been ascribed to the springtime stratospheric O_3 hole over Antarctica which now often covers that region as well.

Further reading

Farman, J.C., Gardiner, B.G. and Shanklin, J.D. *Nature* **315**, 1985, 207–212.

Kirk, J.T.O. *Light and Photosynthesis in Aquatic Ecosystems.* Cambridge University Press, 1982, Cambridge.

Krupa, S.V. and Kickert, R.N. *Environmental Pollution* **61**, 1989, 263–393.

Jones, R.R. and Wigley, T. (eds) *Ozone Depletion: Health and Environmental Consequences.* John Wiley & Sons, 1989, Chichester.

Schell, R.C. et al. *Nature* **351**, 1991, 726–729.

Smith, R.C. et al. *Science* **255**, 1992, 952–959.

Solomon, S. and Albritton, D.L. *Nature* **357**, 1992, 33–37.

UNEP Environmental Effects Panel Reports. *Environmental Effects of Ozone Depletion.* 1989 Report and 1991 Update, United Nations, Nairobi, Kenya.

Worrest, R.C. and Caldwell, M.C. (eds) *Stratospheric Ozone Reduction, Solar Ultraviolet Radiation and Plant Life.* NATO ASI Series G, Ecological Sciences, Springer-Verlag, 1986, Berlin.

Chapter 8

GLOBAL WARMING

'Now is the time for the burning of the leaves'
from 'The Burning of the Leaves' by Laurence Binyon (1869–1943).

The 'greenhouse effect'

A real process

When radiation hits a surface or a gas molecule, it loses energy which causes the wavelength of that radiation to lengthen. This means that when high energy solar radiation (mainly visible and long-wave UV) enters the atmosphere, some is immediately reflected back into space by clouds, etc., but when about half reaches the planetary surface, most of this is reflected back as lower energy infrared (IR) radiation. A small proportion of this (10%) passes directly out into space but the main component is absorbed by certain gaseous molecules in the atmosphere. These molecules then radiate a proportion of this absorbed IR energy in all directions, some out to space and some back to the Earth's surface. Furthermore, the warming of the Earth's surface, directly by solar radiation and indirectly by reradiated IR, causes both evaporation of water and upwards convection of air, both of which transfer energy from the Earth's surface back into the atmosphere. The overall effect is a warming, blanket-like process around the Earth's surface known as the greenhouse effect which is summarized in Fig. 8.1.

There is no doubt that the greenhouse effect is real. The mean global temperature is already 33°C higher now than it would be if no greenhouse gases were present in the atmosphere (i.e. 12°C, not −21°C). Furthermore, the atmospheres of Mars and Venus are very different from those of the Earth but their surface temperatures are exactly those predicted by the greenhouse effect as well.

The greenhouse gases

Details of the important gases are shown in Table 8.1. Some of these exist for long periods in the atmosphere (e.g. CO_2, N_2O and CFCs)

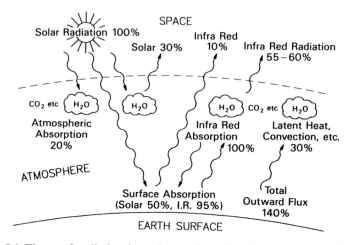

Fig. 8.1 Fluxes of radiation in and out of the Earth's atmosphere which collectively give rise to the greenhouse effect.

Table 8.1   Past, present, and future details of important greenhouse gases excluding water vapour molecules (adapted from IPCC and Deutscher Bundestag data)

Parameter	CO_2	CH_4	N_2O	O_3	CFCs
Concentration (1993)	362 µl l^{-1}	1.75 µl l^{-1}	313 nl l^{-1}	20 nl l^{-1}	0.6 nl l^{-1}
Concentration (1765)	280 µl l^{-1}	0.8 µl l^{-1}	288 nl l^{-1}	15 nl l^{-1}	—
Current % yearly increase	0.45	0.95	0.25	0.5	5
Lifetime (years)	100	10	150	0.1	100
Past radiative forcing (W m^{-2}) for the period 1765–1993	1.5	1.56[a]	0.1	0.05	0.2
Warming potential per molecule ($\equiv$ numbers of CO_2 molecules)	1	32	150	2 000	15 500
Current % contribution to global warming[b]	50	19	4	8	17

[a] Radiative forcing is the difference between total incoming solar radiation (240 W m^{-2}) reaching the surface of the Earth and radiation emitted back out into space. This is the amount of actual global warming. In the case of CH_4, an extra 0.14 W m^{-2} has been added so as to account for additional water vapour arising from the atmospheric oxidation of CH_4.
[b] These percentages exclude water vapour molecules which, being the greatest contributors to the greenhouse effect, currently add an additional 65% to total global warming.

while others are of short duration (e.g. tropospheric O_3). Currently, levels of all of them are rising and, with the exception of CH_4 and CO_2 itself, their sources, sinks, and effects have been covered elsewhere (N_2O, Chapter 3; O_3, Chapter 6; CFCs, Chapter 7). The predominant CO_2 contributes about 50% to overall warming but the others are important

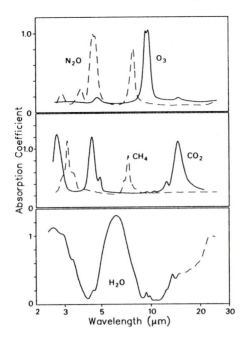

Fig. 8.2. Individual absorption spectra of the different greenhouse gases.

because all of them are more efficient absorbers of IR then CO_2. A molecule of O_3, for example, absorbs 2000 times as much energy as a CO_2 molecule, and a CFC molecule absorbs even more. The greenhouse gases also absorb energy right across the IR spectrum (Fig. 8.2). This means there are virtually no windows through which reflected IR may escape into space.

Methane

Methane (CH_4) levels in the atmosphere are also rising as a result of various human activities. Currently, atmospheric concentrations are around 1.75 µl l^{-1}. Table 8.2 lists the various sources and sinks of CH_4 in terms of their global significance. Most emissions of CH_4 arise from methanogenic bacteria in anaerobic environments but a small proportion is of abiogenic origin from deeper inside the Earth's crust. All methanogens require very low levels of O_2 to form CH_4 but they grow well over a wide range of temperatures and pH levels. They use a variety of substrates as well as H_2 to form CH_4 but the commonest is CO_2. However, CO, formate, acetate and methanol often substitute for CO_2 in different methanogens. The mechanism in all, however, is very similar.

Table 8.2   Estimates of global annual turnover of CH_4 (adapted from various sources)

Natural sources	$Mt\ a^{-1}$	Homogenic sources	$Mt\ a^{-1}$	Sinks	$Mt\ a^{-1}$
Wetlands	120–200	Landfill	30–70	Tropospheric ˙OH oxidation	375–475
Termites	5–150	Biomass burning	30–100	Stratospheric decomposition	35–50
Oceans	7–13	Rice paddies	70–170	Soil absorption	10–30
Ruminants (wild)	2–6	Ruminants (domestic)	70–80		
Lakes	2–6	Gas venting, pipeline leaks and flaring	25–50		
Tundra	1–5	Coal mining	10–35		
Others, e.g. volcanoes	0–80	Sewage operations	5–70		

The substrate (e.g. CO_2) binds to a five-membered carbon ring structure (a furan) and is then progressively reduced through formyl (HCO–), methenyl (CH=), methylene (CH_2), and methyl (CH_3) stages, finally releasing CH_4. If, however, the substrates are formate, acetate, or methanol then they bind to the furan at the appropriate stages.

Of the natural biogenic sources, those from microbes in wetlands are the most important but amounts emitted from methanogens in ocean sediments and termite mounds are also significant. Some concern has also been expressed about future increases in tundra emissions of CH_4. Should global warming become extensive, especially in northern latitudes as the computer models are suggesting, then part of the permafrost could melt and the freshly released wetlands will also start to release additional CH_4.

Levels of CH_4 produced as a result of human activities are three to four times greater than those from natural sources. Of these, current estimates of the amounts of CH_4 released from rice paddies are the greatest and are still being revised upwards as more Chinese information becomes available. The burning of forests and other materials at low temperatures also produces significant quantities of CH_4, as well as CO and CO_2, while emissions from landfills (containing domestic and industrial rubbish) or seepage from natural gas pipelines and coal mines are also important CH_4 sources. Finally, the world's large population of domestic ruminants emit copious amounts of CH_4 from both ends – 10 times as much as from wild animals.

The methanogenic bacteria of ruminants usually produce CO_2 and CH_4 from acetate (CH_3COO^-) but other bacteria are responsible for the breakdown of more complex organics like propionate and butyrate. These *Ruminococci* only start this breakdown to release acetate and H_2

when the methanogens can accept this H_2 for reduction of the acetate to CH_4. This necessary coordination between two different types of bacteria is called syntropy.

The major sink for CH_4 is tropospheric oxidation by ${}^{\bullet}OH$ radicals to form H_2O and CO_2 (see Fig. 6.2). Only a tenth of this amount of CH_4 reaches the stratosphere but, when oxidized, it forms a significant proportion of the stratospheric water vapour.

Water vapour

Although omitted from Table 8.1, Fig. 8.2 also demonstrates that the most important greenhouse component is water vapour which is an IR absorber/radiator both inside and outside clouds. Levels of water vapour, in turn, are highly dependent upon temperature, especially with respect to the capability of air to absorb water vapour. In fact, the contribution of water vapour to the greenhouse effect is a function of the square of the water vapour pressure. Moreover, the flux of latent heat into the atmosphere increases during precipitation. The net effect is to magnify the greenhouse effect at the tropics.

It is also possible that the patterns of monsoon rains could change as a result of global warming. If the land warms more rapidly than the oceans before the monsoons start then the monsoons that follow could be more intense and this could be at the expense of pre-monsoon rainfall which is vital for seed germination.

Past temperatures and carbon dioxide/methane levels

Probable temperatures in the past over temperate regions are shown in Fig. 8.3. By themselves, they provide little indication for future trends, but past records of greenhouse gases such as CO_2 and CH_4 established by measurements taken from ice cores agree with local temperatures which have existed at the poles (Fig. 8.4). These indicate that the large swings ($\pm 3.5°C$) in temperature across the various ice ages and interglacial periods correlate closely with coexisting CO_2 and CH_4 concentrations. However, they still tend to reflect rather than to explain why the Earth moves into an ice age and out again.

Levels of CO_2 in the lower atmosphere (troposphere) are rising year by year (Fig. 8.5). During the last ice age (Fig. 8.4), they were below 200 $\mu l\, l^{-1}$, and they were still around 280 $\mu l\, l^{-1}$ before the Industrial Revolution, but they have risen to 360 $\mu l\, l^{-1}$ since then (see Table 8.1). Over the period 1860–1960, the rise in CO_2 was 9% a^{-1} but, by the year 2000, the increase over the century as a whole will have been 25% a^{-1}.

Most scientists now agree that these increases are probably (rather

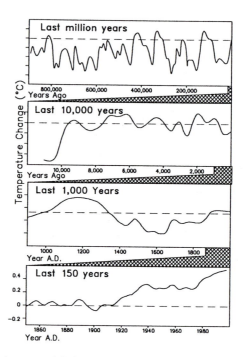

Fig. 8.3 Changes in mean global temperatures in the past over different time scales until the present time. All values of temperature change are expressed relative to that in year 1900 shown as a horizontal dashed line.

than conclusively) due to human activities. By burning large quantities of fossil coal or oil, and destroying tropical rain forests at the rate of 0.5 ha s^{-1} and replacing them with grasslands and waste scrub, human activity has tipped the net balance of the natural Carbon Cycle (Fig. 8.6) in favour of more CO_2 being returned to the atmosphere than at any time since the end of the last ice age (see Fig. 8.4). When expressed in global quantities of carbon exchanged over one year the full extent of this imbalance can be appreciated. The 5 Tt of fossil fuels burned in 1976, for example, contributed an extra 2.3 μl l^{-1} CO_2 to the atmosphere. Of this total, 1.6 μl l^{-1} was either absorbed by the oceans or added to biomass, leaving a net increase of 0.7 μl l^{-1} CO_2 in the atmosphere by the end of 1976.

At ground level, there are considerable local variations in temperature, and there are different temperature gradients upwards through the atmosphere (Fig. 1.3). In the terms of plant growth, the average decrease of 1°C for every 300 m rise in altitude is the most significant but, in global terms, these local variations of temperature make very little difference – mean global temperatures never vary more than a degree

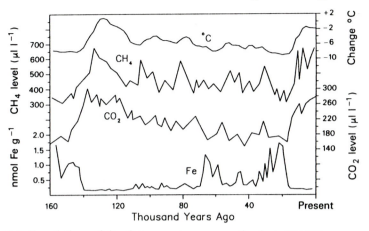

Fig. 8.4 Correlation of local temperatures over the last 160 000 years with analyses of ice core samples from Antarctica (Vostok) for trapped CO_2 and CH_4 as well as with iron deposition.

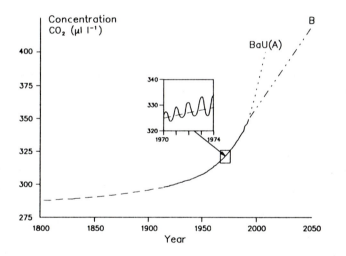

Fig. 8.5 Global changes in atmospheric CO_2 with time. The solid part of the graph represents actual measured values and the dashed regions on the right are those most likely to occur in the future either with 'business-as-usual' – BaU (A) – or if the IPCC predictions of Scenario B (**B**) are brought into reality by international implementation. Typical seasonal fluctuations are shown in the inset figure. Values in the northern summer are lowest because of the higher photosynthetic activity of northern forests.

either side of 12°C. Should this change, even by a degree, then the consequences are likely to be considerable. For example, between 6000 and 8000 years ago, a temperature rise of 2°C was responsible for

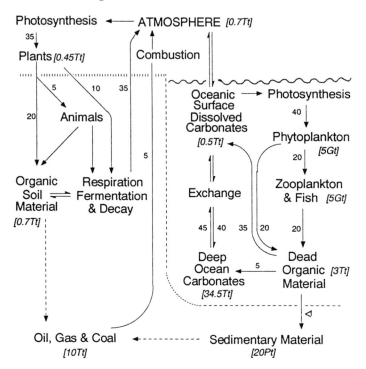

Fig. 8.6 The global carbon cycle where total pool sizes are shown in large type and fluxes in small type. The units of the latter are Gt a^{-1} (for units, see Appendix 3).

melting the ice sheets spread across both hemispheres and brought about the end of the last ice age (Fig. 8.3).

Records of past temperatures on Earth in relation to CO_2 concentrations have been taken back much further than those illustrated in Fig. 8.4. The abundance of ^{13}C in atmospheric CO_2 at present is 1.11% as compared to that of the major isotope (^{12}C), while ^{14}C levels are less than $10^{-10}\%$. In fossil fuels, however, all the ^{14}C has decayed and ^{13}C levels are lower than in the present atmosphere. During photosynthesis, there is a discrimination between the two natural isotopes of carbon where $^{12}CO_2$ is preferred to $^{13}CO_2$ by the CO_2 fixation enzyme, RubisCO. By comparing the $^{13}C/^{12}C$ ratios in planktonic Foraminifera from surface waters of the oceans with benthonic Foraminifera, which occur in deep waters, it is therefore possible to measure the fraction of CO_2 returning from upwelling deep waters that is then photosynthesized. It is also possible to measure the isotopic exchange of carbon along the flow paths of deep waters and the amounts of carbon falling downwards from the organic debris of the surface waters. By deep ocean core sampling and separate analysis of planktonic and benthonic

Foraminifera, it is now possible to take the record back over time well beyond that achieved by polar ice core gas sampling. As a result, the record of likely CO_2 concentrations in deep oceans and in surface waters in contact with former atmospheres extends back over 350 000 years. By comparing these patterns of change with known periods of polar ice retreat and advance, it is clear that rises in atmospheric CO_2 precede global warming.

Future temperatures, sea levels and rainfall

Predicting what is likely to happen in the future is more difficult. It is quite evident from Fig. 8.5 that CO_2 levels are likely to rise exponentially if nothing is done to regulate emissions. This is referred to as Scenario A or 'Business as Usual' (BaU) by the Intergovernmental Panel on Climate Change (IPCC) which is a joint committee set up by the World Meteorological Organization (WMO) and UNEP. Scenarios B–D also proposed by them are a series of progressively more stringent emission control strategies which will require extensive international agreement. Beyond BaU, only Scenario B is remotely realistic in terms of possible international agreement and implementation. Even this assumes that the energy supply mix in all countries moves to lower carbon fuels (i.e. natural gas and nuclear), large fuel efficiencies are obtained, controls on CO emissions are effective, deforestation is reversed, and the Montreal Protocol for CFCs is implemented in full by all countries. Even so, with all these limitations in place, CO_2 emissions will still increase (Scenario B, Fig. 8.5) but less rapidly than with BaU.

The strictures of Scenario B also illustrate that the problem of global warming is not confined to CO_2. All the greenhouse gases have to be taken into account, and if possible limited. Moreover, temperature as such does not provide the means of equating the potential effects of each greenhouse gas. Consideration must be given to total energy inputs to the Earth's surface set against total energy losses. Total incident solar radiation, on average, amounts to 340 W m^{-2} and immediate reflected losses to outer space are around 100 W m^{-2}. The difference between the two (240 W m^{-2}) is the amount absorbed at the Earth's surface. The difference between this surface absorption and the amount of energy emitted as IR and subsequently radiated back from greenhouse gases and water vapour molecules is known as radiative forcing. Each greenhouse gas contributes different proportions to radiative forcing and, because their relative proportions and efficiencies of IR absorption alter with time, their future contributions to the greenhouse effect also change. Figure 8.7 shows the predicted changes in radiative forcing due to CO_2 alone with time, then with added CH_4, and finally with other greenhouse gases (CFCs, N_2O, and tropospheric O_3) added. The values

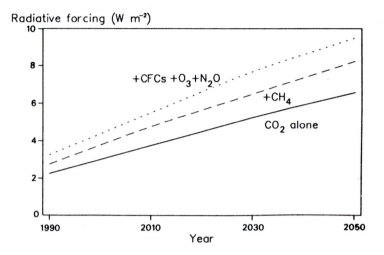

Fig. 8.7 Likely future changes in global radiative forcing under a BaU scenario for CO_2 alone, for CO_2 plus CH_4, or for CO_2, CH_4, CFCs, O_3 and N_2O combined together (IPCC predictions).

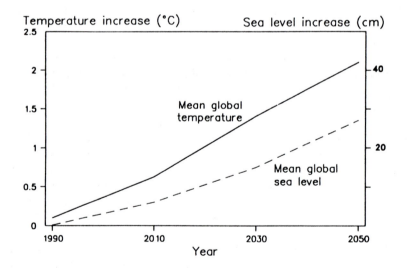

Fig. 8.8 Likely changes in mean global temperatures and sea levels (IPCC predictions).

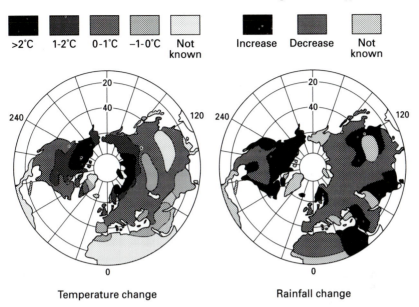

>2°C 1-2°C 0-1°C −1-0°C Not known   Increase Decrease Not known

Temperature change   Rainfall change

Fig. 8.9 Likely changes in future temperature and rainfall patterns across the northern hemisphere.

shown also assume a BaU scenario, but the curve for CO_2 alone would be the same as that for all greenhouse gases if Scenario B was ever implemented effectively.

The major consequences of such radiative forcing evaluations (Fig. 8.7) under a BaU scenario are twofold. Mean global temperatures are likely to rise by 1.4°C (±0.7°) by year 2030, and by 2.1°C (±0.8°C) by year 2050 (see Fig. 8.8) – rather less than had been predicted several years ago. Furthermore, global mean sea level will be 20 cm higher in year 2030 and 31 cm higher in 2050. Most of this increase in sea level comes about from an increase in sea volume due to thermal expansion and rather less from the melting of polar ice. These predictions of sea level rise are also less than those of a few years ago but will still cause extensive flooding of low-lying areas in the world, especially during storms. The countries most at risk are the river delta areas of Bangladesh, Egypt, Thailand, and China (most vulnerable first) but many small islands will also suffer. The immediate effect of inundation will be crop destruction but, more importantly, the surrounding groundwaters will become increasingly saline and large areas of productive land rendered permanently useless.

Since 1932, the rise in sea level has been 64 mm and construction of water reservoirs has reduced further rises in sea level by 32.5 mm. Unfortunately, the future rate of reservoir construction is not likely to have a significant effect on projected sea level increases, although there is no doubt that additional freshwater supplies will be needed as more areas of the world become more arid, prone to erosion, and desertified. Raising mean temperatures by 1–2°C will enhance both net photosynthesis by plants and their water use efficiency (see later) but this is only of advantage in areas which have sufficient rainfall.

Where and how much rain falls over a particular region depends upon a large number of variables and some of these are likely to change as a consequence of global warming. A number of computer-based models of changes in temperature around the world have been extended to include likely changes in rainfall patterns. Figure 8.9 illustrates possible temperature and rainfall changes across the Northern Hemisphere currently suggested by some of these simulations. Most of these models agree that the greatest proportional increases in temperature will occur in northern Canada and Russia and more precipitation will occur in northern Canada, Scandinavia, the eastern Mediterranean, India, and parts of China. Meanwhile, most of the remaining parts of Asia, Europe (apart from Scandinavia), and the USA will become drier.

Albedo, planetary inclination, and volcanoes

Other natural phenomena profoundly affect the greenhouse effect. The albedo or reflectiveness of the planet has a marked effect on the proportions of energy being reflected back into space. As deserts increase in area through a combination of overgrazing, global warming and changes in rainfall pattern, the albedo increases. This has the effect of diminishing the greenhouse effect. Similarly, increases in amounts of dust particles from wind-blow and aerosols from human activities also increase planetary reflectiveness. Melting of polar ice, by contrast, will have the opposite effect.

Changes in the Earth's orbit and volcanic activity are also important but on very different time scales. The Earth shows variations in inclination, precession, and eccentricity in orbit over very long periods. Of these, the cycle of inclination (the angle of the Earth's spin in relation to the Sun), taking 40 000 years, is the most relevant to mean global temperatures. At 24° inclination, more of the poles are presented to incoming radiation than at 22° when the ice caps diminish and the tropics expand. Currently, we are at the midway point (23°, heading towards 22°) in terms of the angle of inclination so significant global cooling from this mechanism is not likely to occur for another 15 000 years or so.

Volcanic eruptions, however, have rapid effects as dusts and gases are hurled up into the stratosphere where SO_2 is rapidly oxidized to SO_4^{2-} particles. Such aerosols have a high albedo which brings about a significant cooling within months. If the volcanic eruption is particularly great, this cooling effect can last for years. The largest eruptions within the industrial period have been Timbora in 1815, Krakatoa in 1883, and El Chinchon in 1982 (see Plate 7). A year after the latter, calculations showed that solar radiation reaching the Earth's surface in the Northern Hemisphere was reduced by 28%. Similarly in 1816, the 'year with no summer', it has since been estimated that global mean temperatures were reduced by more than 0.5°C.

As temperature is directly related to plant growth, dendrochronological studies show an excellent relationship between volcanic activity and periods when tree rings narrow. These marker incidents have allowed different sets of tree ring studies spanning several thousand years to be linked together and have established an alternative method of dating archeological remains.

Oceanic influences

Carbon dioxide is unusual among atmospheric gases in that most of the global atmospheric–oceanic total is dissolved in the oceans (98.5%) and even more is locked away in limestone as a result of previous biological activity in the oceans (Fig. 8.6). By contrast, less than 0.6% of total atmospheric–oceanic O_2 is in the oceans. This difference is due to the fact that CO_2 is 30 times more soluble than O_2 because it reacts with water to form HCO_3^- and CO_3^{2-} while very little ($<1\%$) remains unionized.

There are also marked differences in dissolved CO_2 at the surface as compared to deeper waters (12% less at the surface). This is due to two pumping processes. The solubility pump occurs because CO_2 is more than twice as soluble in cold seawater, either deeper down (or at higher latitudes), than in warmer surface water (or at low latitudes). This solubility pump therefore pushes dissolved CO_2 downwards (or towards the poles). The second pump is known as the biological pump and arises from the photosynthetic activity of the phytoplankton in the euphotic zone. Much of the fixed carbon is cycled in this zone by respiration, etc., but at least 30% sinks into deeper ocean. Meanwhile, other nutrients (especially dissolved iron) in the euphotic zone are heavily depleted and, consequently, their availability (not light, temperatute, CO_2, etc.) becomes the main limitation to growth for phytoplankton over large areas of the oceans.

Calculations using existing HCO_3^- and CO_3^{2-} levels in the oceans have shown that, without both pumps, current levels of atmospheric CO_2 would be 720 µl l^{-1}. If just the solubility pump were to operate, atmospheric CO_2 would then fall to 450 µl l^{-1}. Finally, if the biological pump is then added and all surface nutrients are depleted, the atmospheric level of CO_2 would be 165 µl l^{-1}. Consequently, if small amounts of nutrients are available, a pre-industrial atmospheric level of CO_2 around 280 μl l^{-1} is entirely realistic (Table 8.1).

The major sources of nutrients for growth of phytoplankton are upwellings of colder water from the ocean deeps rather than run-off or wind-blow from land masses. Many fisheries depend on these regular upwellings which, in turn, are affected by surface currents. Fears have been expressed that global warming may influence surface currents which then change the pattern of upwellings. Such changes are very difficult to predict by modelling but some progress is being made.

The most important micronutrient which limits the growth of phytoplankton over most of the oceans is biologically available dissolved iron – rather different from the situation in lakes and shallow sea masses such as the Baltic where phosphate levels are usually more critical. Recently, it has been suggested by J. H. Martin that the availability of dissolved iron from dust deposition in the past may have increased rates of oceanic photosynthesis to such an extent that atmospheric CO_2 levels were much reduced and the resultant global cooling caused the onset of ice ages. Indeed, high levels of iron inversely correlate with low partial pressures of CO_2 in the Vostok ice core samples taken in Antarctica (Fig. 8.4).

This hypothesis has been extended to suggest that artificial fertilization of certain oceans with iron in a biologically available form could bring about a significant reduction of atmospheric CO_2 levels in the future. Unfortunately, calculations show that iron fertilization of the whole of the southern oceans (a huge task) would remove only 2 Gt CO_2 a^{-1} of the current rate of atmospheric CO_2 increase (7 Gt a^{-1}). Moreover, the increase in net oceanic photosynthesis will probably release so much extra N_2O that global warming would get worse (not better) and significant stratospheric O_3 depletion would also occur. Furthermore, emissions of CH_4 and CH_3SCH_3 would also increase and have additional adverse effects on global warming and rainfall patterns.

Plant effects

Some of the greenhouse gases like tropospheric O_3 have direct effects on vegetation (see Chapter 6) while CFCs (Chapter 7) indirectly affect plant growth by enhancing the flux of UV-B radiation. Others, such as CH_4 and N_2O, appear to have no direct effects on plants, although they form significant microbial parts of the carbon and nitrogen cycles which also involve plants, soils and the atmosphere. However, it is CO_2 as a source of photosynthate that dominates thought on likely effects of global warming on plant growth. This, however, is too simplistic because other features of global warming, apart from CO_2, affect plant growth. Temperature, rainfall (water availability), the length and quality of

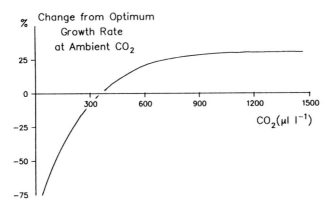

Fig. 8.10 Enriching atmospheres above horticultural crops with CO_2 usually gives improved yields but, above 1000 μl l^{-1}, little improvement is to be gained and a heavy penalty is paid if levels of CO_2 fall below ambient.

incident sunlight, and the availability of other nutrients (e.g. N) are all important determinants of growth and all are likely to change as a result of global warming.

Photosynthesis

Much is already known about plant growth at high levels of atmospheric CO_2 because CO_2 enrichment is often used during commercial horticulture inside glasshouses. Originally, oil-fired burners were used to enrich glasshouses with CO_2 but, after the costs for heating soared, the practice of redirecting flue gases directly into the glasshouses from heating systems (which burn either propane or natural gas) has become more common. The possible benefits in growth that are obtained during commercial horticulture from such CO_2 enrichment are shown by Fig. 8.10. Although there are variations in growth even between cultivars of the same crop species, as a general rule an additional 30% of growth is obtained by raising glasshouse atmospheres to 1000 μl l^{-1} CO_2 – thereafter, the gains are not significant. The additional nitrogen oxide pollution problems, which often abolish all these possible gains, have been covered elsewhere (Chapter 3).

Significant reductions in growth due to deprivation of CO_2 are frequently experienced by plants outdoors. In crops growing close to-gether, such as in tropical and subtropical forest canopies, levels of CO_2 are often reduced well below ambient. Over long periods of time, competition for these reduced amounts of CO_2 between species within canopies favours those species that are better able to capture CO_2 than their neighbours. In subtropical climates, this evolutionary pressure has

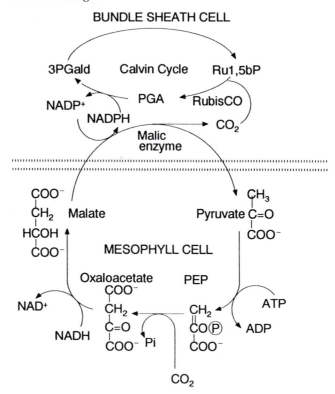

BUNDLE SHEATH CELL

MESOPHYLL CELL

Fig. 8.11 Outline of C_4-type photosynthesis. Three different types of metabolism are known but that shown here is characteristic of maize, sugarcane, and sorghum. Abbreviations: PEP = phospho-enol-pyruvate; PGA = phosphoglycerate; 3PGald = 3-phosphoglyceraldehyde; Ru1,5bP = ribulose-1,5-bis-phosphate.

caused the emergence of so-called C_4 plants with more efficient CO_2 concentrating mechanisms.

All plants trap atmospheric CO_2 by means of the enzyme RubisCO with the assistance of other enzymes of the Calvin (C_3) Cycle. Such a process appears to have existed since the evolution of photosynthesis (about 2.5 billion years ago) but, occasionally since then, a significant modification to conventional C_3 photosynthesis involving four carbon (C_4) acids has occurred in a range of plant species which include some very important crops such as maize, sorghum, sugarcane, and millet. Modern understanding of this modification now would be to regard C_4 photosynthesis not as a replacement for C_3 photosynthesis but as a beneficial adjunct to this basic process under certain environmental conditions. Perhaps the best way to appreciate this modification to photosynthesis is to consider it as being similar to the improvement to the performance of a combustion engine one would gain by adding a turbocharger.

All types of C_4 plants discovered so far have a different arrangement of leaf cells as compared to normal C_3 plants. Outer mesophyll cells completely surround inner bundle sheath cells which in turn enclose the vascular elements leading from the leaves down to the roots. The basic feature of C_4 photosynthesis is the primary assimilation of CO_2 by the carboxylation of phospho-enol-pyruvate (PEP) catalysed by PEP carboxylase (a more efficient CO_2 trapping enzyme than RubisCO; Fig. 8.11). The oxaloacetate so formed (a C_4 compound) is then reduced to malate or transaminated to aspartate and then transported from mesophyll to bundle sheath cells through cell-to-cell connections (plasmodesmata). Subsequent metabolic steps in different C_4 variants differ, but the CO_2 released by decarboxylation is then refixed by RubisCO in the plastids of the bundle sheath cells. The C_3 compounds, pyruvate or alanine, left over are then transferred back into the mesophyll cells where they are converted back into the C_3 precursor (i.e. PEP), which completes the cycle of C_4 photosynthesis across the two types of cell (Fig. 8.11).

The essence of the C_4 process is that advantage is taken of the higher affinity for CO_2 of the enzyme PEP carboxylase over that of RubisCO, an enzyme which is unusual in that it can either carboxylate or oxidize RuBP. The latter alternative is favoured at low internal CO_2 concentrations and is the starting point for a process known as photorespiration which consumes O_2 and releases CO_2 (see Chapter 9). In C_4 plants, the oxygenase activity of RubisCO is suppressed by the CO_2-pumping action of PEP carboxylase which raises the effective concentration of CO_2 in the bundle sheath plastids and favours carboxylation activity by RubisCO over that of oxidation. Any photorespiratory CO_2 that escapes from the bundle sheath cells is then immediately recaptured by PEP carboxylase. As a consequence, C_4 plants show lower internal concentrations of CO_2 but are still more effective than C_3 plants in being able to reduce CO_2 concentrations around a leaf.

Within stands of vegetation, where amounts of CO_2 are limiting, this property of C_4 plants is of considerable advantage over C_3 plants in warm climates. Theoretically, this should mean that when CO_2 levels are raised this advantage of C_4 plants diminishes. This, in turn, is of significance to global crop productivity because many of the major tropical crops are C_4 plants (maize, sorghum, millet and sugarcane) but most C_3 plants (wheat, rice, barley, oats) grow in temperate regions. This also implies that extra CO_2 may benefit temperate crops in the future and switches from maize to wheat and rice in areas where all three are grown may follow. Moreover, 14 of the world's major weeds that affect C_3 plants are C_4 plants. This means C_3 crops should compete better against some of their weeds at higher CO_2 concentrations. However, atmospheric

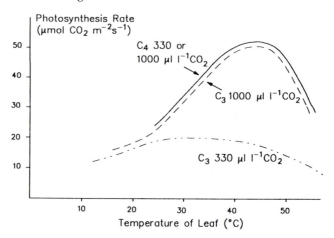

Fig. 8.12 Differences in photosynthetic uptake of C_3 and C_4 plants (*Larrea divaricata* and *Tidestromia oblongifera* respectively) at different temperatures and CO_2 concentrations (adapted from Long & Drake, 1991).

CO_2 concentrations cannot be considered in isolation. Other determinants of plant growth such as temperature must also be taken into account.

Temperature

In the majority of C_3 species, rates of photorespiration (see also Chapter 9) increase faster than the equivalent rates of photosynthesis as temperature rises. Consequently, C_3 plants show a flattened range of temperature optima between 10 and 28°C (Fig. 8.12). In C_4 plants, however, where high internal CO_2 levels preclude the oxygenase activity of RubisCO (i.e. no respiration), the temperature optima of C_4 plants are higher (35–40°C) and more sharply defined. This means that the raising of temperature alone favours plants with C_4 photosynthesis over those using just C_3 mechanisms – exactly the opposite effect of increasing atmospheric CO_2 levels. However, the situation changes as CO_2 levels rise and photorespiration becomes less likely. C_3 plants then tend to behave more like C_4 plants with elevated and sharper temperature optima (Fig. 8.12).

Consequently, only C_3 crops growing beyond their temperature optima are hampered relative to C_4 plants. For most crops, this limitation does not apply – temperature is the major limiting factor for productivity. It controls, for example, the length of the growing season (the time between spring and autumn frosts) which affects a host of growth parameters (e.g. leaf expansion and development, flowering, and fruiting). As a rough guide, a rise in temperature of 1°C lengthens the growing season by 10 days, which means the areas devoted to crops could be extended to higher latitudes and altitudes if the chances of late frosts in spring and early frosts in autumn are diminished. Minimum temperatures are also important for trees and shrubs, especially if they retain their leaves or needles through winter. A general rise in global temperature will again allow the range of the less hardy to extend to higher latitudes and altitudes but this has large consequences for biodiversity. If temperate plants move towards higher latitudes and, at the same time, those plants that currently grow there take advantage of raised temperature and CO_2 then there will be a crushing effect on certain species less well able to compete and they may be eliminated by natural selection.

Water availability

By and large, the consequences of increases in temperature and CO_2 levels taken in isolation appear to be rather encouraging but, unfortunately, water evaporation rates from land surfaces will also increase and cause larger areas to become more arid. Some plants are already adapted to water deprivation, but this major limitation to growth will become more important as the full global consequence of global warming materializes.

Certain succulents and cacti show an alternative type of photosynthesis called crassulacean acid metabolism (CAM) which has many similarities to that found in C_4 plants. The principal differences between them are that malic or aspartic acids are decarboxylated immediately in C_4 plants while in CAM plants they are accumulated during darkness and decarboxylated later the next day. In other words, there is a temporal rather than a spatial separation of the main CO_2-fixing activities of PEP carboxylase and RubisCO. As a consequence, CAM plants lack the double cell cooperation of C_4 plants but have water storing (succulent) tissues instead. This is possible because stomata in CAM plants are normally open at night and close during the day and, consequently, their water use efficiency (i.e. weight of H_2O transpired per weight of CO_2 assimilated) is very high compared with C_3 plants. Usually, CAM photosynthesis is associated with plants that grow in hot, dry habitats with unpredictable rainfall while C_4 plants tend to be midway between CAM plants and C_3 plants in terms of their water use efficiency. This is because the intercellular levels of CO_2 tend to be lower in C_4 plants because their stomata do not have to open quite as wide which means less H_2O is lost by transpiration. A similar situation occurs in C_3 plants when CO_2 levels are raised. Stomata, as a consequence, do not have to open quite as wide to let an equivalent amount of CO_2 in. As a result, less H_2O is lost by transpiration and the water use efficiency in C_3 plants also rises. The drawback to this is that the internal leaf temperature rises

Table 8.3 Mean yield increases determined experimentally for crops and trees grown in atmospheric CO_2 concentrations > 680 µl l^{-1} over those grown at 350 µl l^{-1} CO_2 (adapted from Krupa & Kickert, 1989)

Highest mean yield increase[a]	Agricultural crops or major forestry trees
> 3.0	Cotton
2.5–3.0	Sorghum, okra
2.0–2.5	Grape, eggplant, sweet pepper, Eastern white pine
1.5–2.0	Peas, sweet potato, beans, radish, barley, sugarbeet, Swiss chard, potato, lettuce, alfalfa, soya, maize, tomato, Scots pine, Norway spruce
1.0–1.5	Oats, wheat, fescue grass, rice, strawberry, cabbage, sunflower, endive, clover, roses and other flowers, Douglas fir, apple, loblolly pine, birch

[a] Relative to controls (set at 1.00) grown at 350 µl l^{-} CO_2.

and if this exceeds the temperature optimum then growth and leaf expansion will slow down.

It is sometimes forgotten that plants have experienced significant global warming and cooling several times during their evolution. Therefore, it is to be expected that plant response in terms of adapting to rising atmospheric levels of CO_2 is complex and several adaptations, morphological as well as physiological, may occur. Examination of old herbarium specimens has indicated, for example, that stomatal frequency per unit area of leaf has declined in recent times. The old herbarium specimens also have lower C to N ratios. In other words, the extra C from CO_2 is not balanced by extra N uptake by crops growing now. This has many implications. For example, protein content of seeds may be reduced in proportion to carbohydrate reserves and this in turn may affect the quality, rather than the quantity, of certain cereal-based foodstuffs.

It is adaptations like these that long-term exposures of a wide range of crops, weeds, trees, and natural vegetation are now seeking to define alongside measurements of yield, biomass, etc. The problem is not a simple one of just testing responses at high levels of CO_2 (600–700 µl l^{-1}) against present levels (350 µl l^{-1}), although a large number of experiments in the late 1980s did just that. What is clearly required is definitive experimentation which studies the interactions between CO_2, temperature, water availability and, possibly, O_3 because this will rise as quickly as CO_2 in the next few decades. Unfortunately, even data from two-way interactions are still very limited and current models/predictions of global crop yield have had to use the earlier high-to-low CO_2 information (e.g. Table 8.3) and to impose estimates of the other influences as thought appropriate. Table 8.4 indicates likely global changes in plant productivity following these principles. More definitive interactive assessments are still in progress.

Table 8.4 Likely increases and decreases in overall crop productivity for different regions as estimated by the Canadian Atmospheric Environment Service (after Smit *et al.*, 1989)

Region	Wheat	Maize	Barley	Oats	Soya	Rice
Canada	↑	↑	↓	↓	↓	
USA and Mexico	↓	↓			↓	
South America	↓	↓			↓	
Europe	↓	↓	↓	↓		
Africa	↓					
Russia and Ukraine	↑	↑	↓	↓		
India, China and SE Asia	↓					↑
Australia, New Zealand	↓					

Human implications

The implications of global warming for humans are primarily socio-logical, economic and geopolitical rather than biological. Possible movements of rainfall, mean temperatures, crops, etc., and the displacement of populations due to flooding, drought, and changed economic circumstances are the cause of much current concern and study. Without doubt these are directly linked to the changes in biological processes described above. However, the major concern of predictive studies in the sociological, economic, and political areas must link future changes in population to likely energy requirements.

Figure 8.13 shows the trends in population from the recent past well into the next millennium. The population in the developed world will be relatively stable and that of China will gradually stabilize. Trends elsewhere, especially in India, the rest of Asia and in Africa, however, are likely to be ever upwards. Likely future rates of energy consumption are in a similar direction (Fig. 8.14) although the patterns of different energy sources will change. Increasing use of coal and natural gas is predicted throughout the next century, so there is little or no prospect of holding back CO_2 emissions to 1990 levels unless nuclear and renewable possibilities are scaled up dramatically.

Of most concern is the composite of both figures – the energy requirement per person which is also set to rise (Fig. 8.15). The developed world accounts for most of the current total global energy requirements. If these are held to 1990 levels by international agreement then the developing world's natural aspirations will still cause increases both in total consumption and on a *per capita* basis. The developed world may well be able to improve their energy use efficiency and diversify into alternative and renewable fuel sources. However, unless there is a rapid and sustained transfer of this energy saving technology from developed to

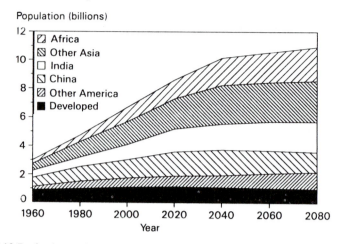

Fig. 8.13 Projections of population increase in different parts of the world (data from the UN Population Division).

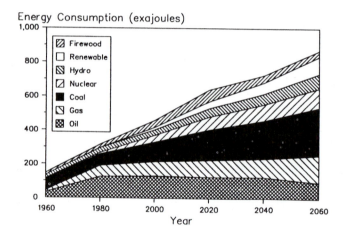

Fig. 8.14 Projected worldwide energy consumption for different types of fuels (data from the World Energy Conference, Cannes, 1986).

developing countries at little or no cost the trend of energy use per person cannot be reversed and global warming, with its attendant problems, will continue.

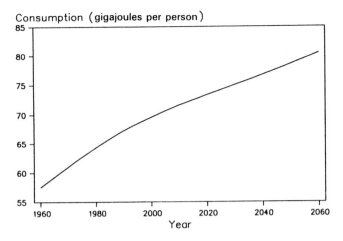

Fig. 8.15 Future likely trend of global energy consumption *per capita* derived from data in Figs. 8.13 and 8.14.

Further reading

Gregory, S. (ed.) *Recent Climatic Change.* Belhaven Press, 1988, London.

Gribbin, J. *Hothouse Earth: The Greenhouse Effect and Gaia.* Transworld Publishers, 1990, London.

Houghton, J.T., Jenkins, G.J. and Ephraums, J.J. (eds) *Climate Change: The IPCC Scientific Assessment.* Cambridge University Press, 1990, Cambridge.

Jäger, J. and Ferguson, H.L. (eds) *Climate Change: Science, Impacts and Policy.* Cambridge University Press, 1991, Cambridge.

Krupa, S.V. and Kickert, R.N. *Environmental Pollution* **61**, 1989, 263–393.

Long, S.P. and Drake, B.G. *Photosynthetic CO_2 assimilation and rising atmospheric CO_2 concentrations.* Ecological Applications **1**, 1991, 129–156.

Martin, J.H. *Paleoceanography* **5**, 1990, 1–13.

Parry, M. *Climate Change and World Agriculture.* Earthscan Publications, 1990, London.

Rogers, J.E. and Whitman W.B. (eds) *Microbial Production and Consumption of Greenhouse Gases: Methane, Nitrogen Oxides, and Halomethanes.* American Society of Microbiology, 1991, Washington, DC.

Smit, B. *et al. Climatic Change* **14**, 1989, 153–174.

Woodward, F.I. Predicting plant responses to global environmental change. *New Phytologist* **122**, 1992, 239–251.

Wyman, R.L. (ed.) *Global Climate Change and Life on Earth.* Routledge, Chapman & Hall, 1991, New York and London.

Chapter 9

OTHER POLLUTANTS – GLOBAL AND LOCAL

'Youk'n hide de fier, but w'at you gwine to do wid de smoke?'
from 'Uncle Remus – Plantation Proverbs' by Joel Chandler Harris (1848–1908).

Harmful air pollutants

A wide variety of pollutants occur in the atmosphere from time to time. Many of the major ones that affect both plants and animals have already been discussed in earlier chapters. There are, however, a number of pollutants that often occur in certain working environments or in neighbourhoods close to certain industries which are especially harmful to humans. Table 9.1 lists the various clinical conditions which may be induced or aggravated by a wide range of air pollutants. Space in this book does not permit full coverage of these, but important examples (printed in **bold type**) are discussed in further detail below to illustrate the diversity of effects on human (and plant) health caused by some of them.

Oxygen versus carbon dioxide

Co-evolution of the atmosphere and the biosphere

Although vital for many living systems, O_2 encourages a number of free radical reactions within living cells and, as a consequence, is still one of the major problems encountered by biological systems. A pollutant is sometimes defined as 'a chemical in the wrong place and at the wrong concentration'. This means that, in planetary and stellar terms, present levels of O_2 in our atmosphere may qualify O_2 as a pollutant. Perhaps 'at the wrong time' should be added to disqualify O_2. Current atmospheric levels of O_2 around 21% are midway between 16% O_2 where combustion is not possible and 26% O_2 where spontaneous combustion would take place.

During the formative years of this planet, the atmosphere mainly consisted of CO_2. Free atmospheric O_2 only appeared as photosynthetic organisms acquired the ability to split H_2O into O_2, electrons, and H^+

Table 9.1 Clinical symptoms induced by large quantities of airborne pollutants (adapted and updated from Waldbott, 1978)
(Entries in **bold type** are discussed in detail in this chapter)

Symptom	Pollutant	
	Particulate	*Gaseous*
Airway and lung irritation	Asbestos, Co, C, FeO_2, quartz and silica dusts	SO_2, NO_2, O_3, Cl_2, NH_3
Alopecia (loss of hair)	As, **Pb**	
Anaemia	Mo, **Pb**, Se, V	
Arteriosclerosis and coronary heart disease	Ba, Cd, organophosphates	**CO**, H_2S
Arthritis	F^-	**HF**
Asphyxiation		**CO**, H_2S
Asthma and allergies	Co, pollen, fungi, insect scales, house dust, isocyanates, PVC, epoxy resins, etc.	SO_2, O_3, NO_2, **HCHO**
Bone diseases	Cd, Sr, F^-	**HF**
Brain impairments	B, Hg, **Pb**, Zn, DDT	**CO**, borane
Bronchitis		SO_2, O_3, NO_2
Cancer[a]	As, Be, Cd, Co, Cr, Ni, **Pb**, Se, Sr, asbestos	Benzene, **HCHO**, etc., **radon**
Cyanosis (blue skin tint)	Nitrites	**CO**
Dental caries	Se	
Dental discolorations	F^-	**HF**
Dermatitis	As, Cr, Ni, Se, V, organophosphates	**HCHO**
Diarrhoea	As, **Pb**, F^-	**HF**
Emphysema	Cd	SO_2, NO_2
Eye irritation	V	PANs, **HCHO**, H_2S, NH_3, SO_2, benzyl chloride, acrolein
Fume fevers	Heated B, Mn, Ni, Pb, Sb, Sn, Zn, PTFE	**HF**
Gastroenteritis	As, Hg, **Pb**, Se, Zn, F^-	**CO**, **HF**
Headaches	**Pb**, F^-	
Irritability	**Pb**	
Jaundice	**Pb**, Se, Ti	
Kidney malfunctions	Cd, Hg, **Pb**, Se, various insecticides[b]	Chlorocarbons
Liver malfunctions	Mo, Se, insecticides[b]	**Radon**
Lung fibrosis	Ba, Co, Fe, Ni, Se	
Lung silicosis	Silica and other dusts	
Melanosis (darkened skin)	As	
Muscular impairments	Hg, F^-	**HF**
Nasal irritation	As, Cr, Ni, Se, V, F^-	SO_2, **HF**, Cl_2
Nerve impairments and ataxia (irregular movements)	Mn, Hg, **Pb**	
Parathyroid disturbances	F^-	**HF**
Premature ageing		O_3, PANs
Reproductive problems	Cd, Hg, **Pb**	CS_2
Sarcoidosis (granuloma formation in lungs)	Be, Mn, Zn, PVPs, talcum	
Sleeplessness	**Pb**	
Thyroid disturbances	Ba, Co	
Visual impairment	Se, F^-	PANs, **HF**, CS_2, Cl_2

[a] The list of confirmed and probable hydrocarbon human carcinogens is lengthening. It includes acrylonitrile, epoxy-ethane and propane, dimethyl sulphate, ethylene oxide, epichlorohydrin and benzopyrene. The list of possible but as yet unconfirmed carcinogens is much longer (see Tomatis, 1990). [b] E.g. DDT, the aldrins and lindane.

ions. At the same time, CO_2 levels declined rapidly as CO_2 was fixed by photosynthetic organisms. These carbonaceous products passed into food chains and eventually ended up as carbonates in mollusc shells and corals. The consequence today is that most of the CO_2 from the original atmosphere of the Earth is now locked away in the limestone reserves of the world. Meanwhile, O_2 (the most abundant element of the Earth at 53.8%) has been steadily released into the atmosphere to become toxic or, in most cases, lethal to many primitive organisms because of harmful oxidations within their cells. Those that developed protective mechanisms against O_2 survived but those that failed either died or withdrew to environments where O_2 concentrations were lower. Antecedents of the latter are represented today by a wide array of anaerobic organisms ranging from strict anaerobes that cannot tolerate O_2, through others which survive exposure for some time but do not grow in the presence of O_2, to those that grow in concentrations of up to 10% O_2. Undisturbed or stagnant soils, polluted waters, decaying vegetation, wounds, and the internal passages of animals (especially the mouth and the colon) are all locations where anaerobes flourish.

One reason for the sensitivity of certain organisms to O_2 is that certain enzymes are strongly inhibited by oxidation at their active sites. A good example of this occurs with the nitrogenase complex which fixes atmospheric N_2. Blue-green algae (cyanobacteria) carry out N_2 fixation (see Chapter 3) in separate thick-walled cells (heterocysts) which exclude O_2. Similarly, the root nodules of legumes which contain N_2-fixing bacteria (*Rhizobium*) have leghaemoglobin to bind any O_2 that may appear.

All aerobic organisms have extensive free radical scavenging systems (e.g. vitamin E, glutathione reductase, etc. – see Chapter 6) to protect themselves from oxidants arising from O_2 and many use O_2 as a terminal acceptor for respiration.

Photorespiration in plants

All atmospheric O_2 is or has been produced by photosynthesis and the evolution of life on this planet is intimately linked to the appearance of photosynthetic organisms nearly 3 billion years ago. Even one billion years ago, atmospheric levels of O_2 were less than 1%. Consequently, the majority of plant processes evolved when atmospheric O_2 levels were much lower than now. It is therefore not surprising that rates of fixation of CO_2 by plants may be improved by a factor of 50% or more by reducing O_2 levels below 21%, but if higher concentrations of O_2 than ambient are experienced severe reductions of photosynthesis occur (Fig. 9.1). Inhibited root growth, poor seed viability, and stunted shoot development may also take place because additional free radicals give

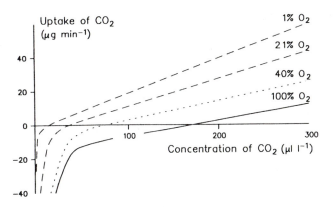

Fig. 9.1 Influence of O_2 upon uptake of CO_2 by plants at various concentrations of CO_2. Positive assimilations represent net fixation of CO_2 into sugars and carbohydrates whilst negative balances represent higher rates of respiration and photorespiration. The intercepts on the horizontal axis (where imports of CO_2 equal exports of CO_2) are known as compensation points.

rise to uncontrolled damage in cellular membranes. Intercepts on the horizontal axis of Fig. 9.1 mark the balance where CO_2 fixation matches respiratory evolution of CO_2. These balances are known as the compensation points and vary with the partial pressures of O_2.

Only part of the evolution of CO_2 by plants is independent of light and due to respiratory activity. The light-dependent remainder is due to a process known as photorespiration which arises from the fact that the active site of the CO_2-trapping enzyme RubisCO may carry out either carboxylation or oxygenation. Subsequent evolution of CO_2 by photo-respiration comes about after further metabolism of the oxygenated products arising from the oxidative mode of RubisCO. On first impression, this appears to be a futile process which throws away any useful CO_2 that might have been fixed. However, photorespiration is now considered to be a vital protective device which assists the dissipation of light energy when amounts of CO_2 are limiting relative to O_2. Normally, such conditions would favour the production of harmful free radicals such as $\cdot O_2^-$ but, by recycling carbon compounds, photorespiration enables the retention of higher internal CO_2 concentrations so that the occurrences of these potential situations are minimized.

Hyperoxia

Possible toxic effects of raised levels of O_2 upon humans have been much studied. Medically, concentrations greater than ambient are used to treat certain cancers, gangrene, multiple sclerosis, and lung ailments. Elevated levels of O_2 are also used in activities such as diving, climbing, flying, and space travel. This has led to a large number of experiments to evaluate, if possible, the short- and long-term consequences of elevated O_2 concentrations at high or low atmospheric pressures. Sometimes, mitochondrial damage due to excess $\cdot O_2{}^-$ formation in the presence of elevated levels of O_2 has been detected in tissues or organs that work hard (e.g. hearts and livers) or produce fresh cells (e.g. bone marrows), but the effects of elevated O_2 vary greatly between species, their ages, the organs or tissue involved, and the diet.

Asphyxiation

The fact that animals use carbon dioxide in a manner different from plants has been known ever since Priestley did his experiments on combustion and respiration in the 1770s. When he burned a candle in an enclosed space, the air inside would no longer support burning and mice died in the residual air but plants lived on for several weeks. After this period, a candle would burn or mice could breathe the air that was left. He correctly concluded that plants used and changed air in a manner different from animals.

The presence of other gases in confined spaces and lack of the ability to breathe under certain conditions had been recognized by miners long before Priestley did his experiments. 'Black Damp', as they called it, consists of air with more than 13% CO_2 and was tested for in a variety of ways. Lighted candles or tapers are extinguished at concentrations of less than 17% O_2, or when CO_2 levels rise to 10%, but this had unfortunate repercussions for the early miners if CH_4 and other explosive gases were also present. Developments of safety lamps improved this situation, but if acetylene was used the flame was not extinguished until levels of 29% of CO_2 had been achieved. A much better safeguard was to take mice along and lower them into any suspect shaft.

Nowadays, CO_2 is widely used in industry and likely to be experienced in relatively high concentration by those who work in many environments (not just mines) unless precautions are taken. This is especially important in those processes which use CO_2 as a refrigerant, as a weak acid for the manufacture of drugs or white lead, in water treatment, whilst filling fire extinguishers, or during food preservation, welding, purging pipelines and carbonated drink production. Sealed systems such as submarines or spacecraft involve perhaps the greatest potential exposure of humans to CO_2 but controls are also necessary in industries such as brewing and wine making. During sugarbeet processing, for example, the juice is treated with lime and the calcium saccharide is precipitated by CO_2. All these operations and industries have produced their fatal accidents from time to time. The most bizarre was the death of four Irish salvage workers who tried to remove a cargo of rotting apples from the wrecked liner *Celtic* so that it could be floated off the rocks outside Cork harbour.

All investigations have shown that it is the deprivation of O_2 that is the cause of harm. Breathing is very deep and rapid at 12% O_2 and severe headaches are experienced. Unconsciousness sets in below this level and death occurs when levels of O_2 are between 8% and 5%. The physiological effect of elevated CO_2 is to stimulate the respiratory centre of the brain near the top of the spinal cord. For example, 5% CO_2 in air produces strong stimulation and is used medically for the resuscitation of gas victims or inducing respiration in newborn babies. Studies in submarines have also shown that atmospheres containing 20% CO_2 may be tolerated for long periods with little effect provided that O_2 levels are adequate. At 30% CO_2, blood pressure increases and hearing becomes difficult, while 30 minutes at 50% produces signs of intoxication. Unconsciousness is produced by 70% CO_2 within a very short period. The TLV of 5% CO_2 for 8 h has a wide margin of safety.

Carbon monoxide

Sources and sinks

After CO_2, carbon monoxide (CO) is the most abundant atmospheric pollutant in the troposphere. Homogenic emissions of CO exceed those of all other pollutants combined and amount globally to 0.75 Gt a^{-1} – an enormous rate and still rising. Emissions are concentrated in the northern hemisphere and show distinctive rhythms – daily, weekly and seasonal – depending upon human activity. Levels of atmospheric CO are highest during winter seasons with the differences between the seasons becoming more marked as time passes. About heavy traffic, CO levels are often well above 50 µl l^{-1} and may exceed 100 µl l^{-1} for long periods in tunnels, garages, and loading bays. Indoors, when smoking is permitted, levels of CO can also exceed 15 µl l^{-1}. Biogenic sources, by contrast, produce a tenth of that produced by human activities. Half of this comes from the oceans and the other half from a variety of natural processes such as volcanoes, electrical storms, and 'marsh gas' from rotting vegetation, etc.

The oceans are a major sink for CO_2 (Chapter 8) but not for CO. A function of partial pressure, the solubility of CO in seawater is much less than that of CO_2 or SO_2. However, the surface waters of oceans contain many times the theoretical CO concentrations that would be expected from gaseous uptake from the atmosphere. In other words, marine biological activity (mainly planktonic) contributes substantial quantities of CO to surface waters and exchanges more CO into the atmosphere.

Over land areas and away from major sources, atmospheric levels of CO are relatively constant which means that natural sinks exist to remove excess CO. Various removal possibilities include oxidation of CO to CO_2 in the troposphere by ˙OH radicals, oxidation in the stratosphere, or absorption by soil microbes or plants.

CO is produced by incomplete combustion and adds to global warming (see Chapter 8) because it is rapidly oxidized to CO_2 (Reaction 9.1) but other reactions with O_2 or nitrogen oxides are less important. Removal of CO, therefore, depends on the presence of sufficient ˙OH radicals which, in turn, requires atomic O and H_2O (Reaction 6.7). Consequently, sunny but wet atmospheric conditions are most favourable for ˙OH production and CO removal – overcast and cold but dry winter days are far from ideal. Any CO that rises above the troposphere into the stratosphere is oxidized to CO_2 by atomic O (Reaction 9.2) or O_3 but this is counterbalanced by photolysis of CO_2 back to CO (Reaction 9.3).

$$\text{˙OH} + CO \Rightarrow CO_2 + H \qquad\qquad (9.1)$$

$$CO + O + M^1 \Rightarrow CO_2 + M \qquad\qquad (9.2)$$

$$CO_2 + light \Rightarrow CO + O \qquad\qquad (9.3)$$

Soils have a variety of soil microbes which oxidize CO to CO_2 or reduce it to CH_4. For example, an anaerobic CH_4 producer like *Methanobacterium formicum* generates H_2 (Reaction 9.4) which then reduces CO or CO_2 (Reactions 9.5 and 9.6). Generally, different soils remove CO at different rates but cultivated soils are much less efficient in doing this than natural soils high in organic matter. Agricultural processes, therefore, select against those microbes most efficient at CO removal. Consequently, the health of non-cultivated soils (e.g. in forests) is vital, not just for protection against acid deposition (see Chapter 5), but also to reduce CO levels.

$$CO + H_2O \Rightarrow CO_2 + H_2 \qquad\qquad (9.4)$$

$$3H_2 + CO \Rightarrow CH_4 + H_2O \qquad\qquad (9.5)$$

$$4H_2 + CO_2 \Rightarrow CH_4 + 2H_2O \qquad\qquad (9.6)$$

Some global estimates of the total potential capacity for soil microbial uptake of CO have been put as high as 14 Gt a^{-1} – well in excess of what is discharged by human activities. By contrast, other calculations have estimated global rates of removal to be 0.45 Gt a^{-1} – much lower than

[1] See Reactions 2.2 and 2.7.

the annual rate of production. Various assumptions account for these discrepancies. Soil microbial uptake has been established as the major global mechanism for CO removal, but this is a slow temperature-dependent process which explains the rhythmical seasonal changes in levels of global atmospheric CO. As northern forest soils warm in summer they supplement the steady uptake of CO by tropical forest soils. Furthermore, as more tropical and temperate areas are brought into cultivation, the cyclical seasonal changes in global CO concentrations become more pronounced. Should northern forest soils also suffer damage by other types of pollution (Chapters 5 and 10) then overall levels of CO and, thereby, CO_2 will rise and enhance global warming.

Plant uptake

Oxidation of CO to CO_2 by certain crops has been detected using ^{14}CO. However, if respiration is artificially inhibited with dinitrophenol then this process is much reduced. The implication of this is that major oxidation of CO by plants is linked to the cytochrome oxidase activity of mitochondria. Some trees, however, do not take up CO at all while green algae have their nitrogen metabolism inhibited by CO. Some seaweeds are even capable of emitting CO. Consequently, any judgement on the relative significance of vegatation to remove CO is difficult to make. Current estimates indicate global uptake of CO by vegetation to be 25% of that of soils.

The oldest industrial poison

Ever since humans learned how to break stones by heat or to burn wood with a restricted flow of air to produce charcoal they have been affected by CO. While smoking is the single most important source of CO for humans, most of the hazards at work connected with CO come from motor vehicles. Blasting, fire fighting, methanol production, wood distillation, and even cooking over charcoal in French restaurants (an affliction known to French doctors as *'Folie des cuisiniers'*) have similar hazards. If combustion processes are improperly ventilated (e.g. during smelting), similar problems arise. Any burner which has a surface cooler than the ignition temperature of the gas phase of the flame starts to form CO. In water heaters, where coils containing water cannot rise above 100°C, large quantities of CO are often given off. This becomes a bigger problem in the home, especially the kitchen, as draughts are restricted to curtail the costs of space heating.

Exposure of humans to CO may cause a variety of clinical effects (Table 9.2). Generally, there are positive relationships between symptoms

Table 9.2   Effects of CO on humans and the accompanying carboxyhaemoglobin blood (CoHb) concentrations

Exposure[a] ($\mu l\ l^{-1}$)	Effects and symptoms	CoHb/Hb[b] (%)
0–10	No discomfort or effect	0–2
10–50	Some tiredness, impaired vigilance and reduction in manual dexterity	2–10
50–100	Slight headache, tiredness and irritability	10–20
100–200	Mild headache	20–30
200–400	Severe headache, visual impairment, nausea, general weakness and vomiting	30–40
400–600	As above, but with greater possibility of collapse	40–50
600–800	Fainting, increased pulse rate and convulsions	50–60
800–1600	Coma, weak pulse and possibility of death	60–70
1600+	Death within a short period	70+

[a] After 2 h exposure.
[b] Likely percentage of carboxyhaemoglobin (CoHb) to total haemoglobin (Hb) contents of blood for some time afterwards in a resting individual but which may be achieved three times faster during heavy work.

and CO levels but the onset of responses is hastened by heat, humidity and exertion. At work, it is the younger rather than the older workers who are more susceptible to CO, but alcoholism, obesity, old age, heart conditions, and lung diseases all exacerbate the problem. However, some physiological compensation or conditioning (e.g. formation of more RBCs) occurs with repeated exposures in steel workers and smokers.

There are clinical differences between O_2 deprivation (hypoxia or anoxia) and CO poisoning. In the former, respiratory symptoms precede nervous symptoms, but in the latter this order is reversed. This has led to the suggestion that the driving out of O_2 from RBCs (see later) is not the only physiological effect of CO.

The progressive symptoms listed in Table 9.2 approach without warning because CO is non-irritating and without smell. Only the slightest presence of other contaminants such as hydrocarbons, mercaptans, ammonia or traces of other materials give an advanced warning of any hazard within fumes. It was a long time before CO poisoning became universally accepted as a specific disease in the industrial and urban environment because many of the early symptoms caused by CO are also associated with repetitive work or stressful situations (e.g. driving). It is now recognized that misjudgements due to enhanced blood CO levels lead to more accidents at work, on the road, and in the home.

Related to CO poisoning is the disease Shinshu myocardiosis. This occurs in those exposed to high levels of CO over long periods and is typified by defective heart valves as well as by angina pectoris and arteriosclerosis. It was first detected in 1955 in the village of Kinasa, Japan where, during cold winters, it had been the custom for silk production to continue in tightly sealed, heated rooms. As a result, many of the workers experienced severe circulatory problems due to CO.

Blood biochemistry and pollution

Haemoglobin, the red pigment of RBCs, consists of the protein globin and a tetrapyrrole ring system called haem which contains ferrous iron (Fe^{2+}). If this iron is oxidized to the ferric form (Fe^{3+}), the haemoglobin turns into methaemoglobin which is unable to bind O_2. Oxidation of haemoglobin to methaemoglobin occurs naturally as RBCs age and is accelerated by the additional presence of H_2O_2. Normally, free radical scavenging (Chapter 6) ensures that this does not occur prematurely but free radical generation by other pollutants such as O_3 or PAN promotes early formation of methaemoglobin. Ultimately, these older RBCs high in methaemoglobin are broken up by certain white blood cells (phagocytes) while they pass through the spleen.

During normal respiration, O_2 combines with haemoglobin to form oxyhaemoglobin – an association–dissociation reaction which is dependent upon the partial pressure of O_2, the temperature, and pH. Changes of blood acidity are largely due to alteration in blood levels of dissolved CO_2 and the equilibrium between the different forms of CO_2 (Reaction 9.7). The enzyme carbonic anhydrase brings about a faster interconversion between dissolved CO_2 and carbonic acid while H^+ ions from the dissociation of carbonic acid cause the dissociation of oxyhaemoglobin.

$$CO_2 + H_2O \Leftrightarrow H_2CO_3 \Leftrightarrow H^+ + HCO_3^- \qquad (9.7)$$

Lactate, which is produced by contracting muscles working at low O_2 tension, also produces acidity which then causes more O_2 to be released from oxyhaemoglobin to alleviate the local shortage. The final factor that encourages the release of O_2 from haemoglobin is 2,3-diphospho-glycerate. This substance combines reversibly with haemoglobin to change its structure in favour of O_2 release. Consequently, 2,3-diphospho-glycerate is released by tissues which are especially short of O_2 to ensure that O_2 is delivered to those cells most in need.

About 7% of CO_2 produced by body tissues is dissolved in the blood plasma and carried to the lungs as a dissolved gas. The bulk (70%) is transported as bicarbonate and the remainder (23%) reacts with haemoglobin to form carbaminohaemoglobin. At the high partial pressures of CO_2 in the tissue capillaries, the formation of this complex is encouraged, but in the lungs where partial pressures of CO_2 are low it splits apart again.

CO combines with haemoglobin with an affinity which is between 200 and 300 times greater than the affinity of haemoglobin for O_2. The resultant carboxyhaemoglobin is extremely stable. CO at very low concentrations in the blood (i.e. 0.1%) will combine with over half of the haemoglobin and immediately reduce the O_2-carrying capacity by a similar proportion. Only high O_2 partial pressures such as those achieved when treating a CO poisoning case with pure O_2 are sufficient to reverse the combination of haemoglobin with CO. Other O_2-releasing mechanisms may also be affected adversely by CO. For example, levels of 2,3-diphospho-glycerate in plasma circulating through O_2-deficient tissues are lower in the presence of CO.

Unfortunately, the affinity of CO for human foetal haemoglobin is higher than that for normal haemoglobin. This means that unborn babies in the womb are especially sensitive to CO poisoning. Undoubtedly, the single most important factor which exposes an unborn child to greater than average CO concentrations is smoking by the mother. The relationship between maternal smoking and low birthweight is undeniable. Most of the number of developmental and clinical defects in the newborn ascribed to deprivation of O_2 are mostly neurological. However, as with adults, it is still not possible to distinguish in these babies the direct effects of CO caused by O_2 deprivation from those caused by other factors in tobacco smoke.

The WHO recommends that levels of CO are reduced so that the level of carboxyhaemoglobin in the blood of the general population does not exceed 3%. This means that exposures should not exceed 50 µl l^{-1} over 30 minutes, 25 µl l^{-1} over 1 h, and 10 µl l^{-1} over 8–24 h.

Formaldehyde

Sources and uses

Formaldehyde (HCHO) occurs in the outside atmosphere at concentrations around 0.1–1 nl l^{-1}. The major outdoor source is the oxidation of methane (see Fig. 6.2) to the methoxy ($CH_3O^{\cdot}$) radical which then reacts with O_2 to form HCHO and $HO_2^{\cdot}$ radicals (Reaction 9.8). HCHO may then react with any available $^{\cdot}$OH radicals (Reaction 9.9) or undergo photolysis which releases either atomic H and the formyl radical (HCO$^{\cdot}$). This then reacts with O_2 to form $HO_2^{\cdot}$ radicals and CO (Reactions 9.10 and 9.11) or CO and molecular H_2 directly (Reaction 9.12).

$$CH_3O^{\bullet} + O_2 \Rightarrow HCHO + HO_2^{\bullet} \tag{9.8}$$

$$HCHO + {}^{\bullet}OH \Rightarrow HCO^{\bullet} + H_2O \tag{9.9}$$

$$HCHO + light \Rightarrow HCO^{\bullet} + H \tag{9.10}$$

$$HCO^{\bullet} + O_2 \Rightarrow HO_2^{\bullet} + CO \tag{9.11}$$

$$HCHO + light \Rightarrow CO + H_2 \tag{9.12}$$

Soil and water surfaces are alternative sinks for HCHO. As a result, HCHO exists as a trace gas for some time (mean residence time, 5–10 days) depending on levels of sunlight, other air pollutants, and various climatic conditions.

Of greater concern from a human health standpoint are indoor levels of HCHO. Many modern buildings use chipboard made with HCHO-based resins or have urea-HCHO-based insulation which slowly releases HCHO gas. Often, this contributes to 'sick-building syndrome'. Furthermore, HCHO is frequently employed as a corrosion inhibitor or as a hardening or reducing agent, as well as in preservatives, embalming fluids, or sterilizing solutions. As a result, indoor levels of HCHO can exceed $1 \, \mu l \, l^{-1}$ and even more if smoking is permitted.

Health hazards

HCHO irritates the eyes, nose and throat of most adults at concentrations above $0.1 \, \mu l \, l^{-1}$. At higher concentrations ($5 \, \mu l \, l^{-1}$), coughing and chest tightness are very evident. If these levels persist or rise, more permanent injury may occur to linings of the airways and the lung tissues. Clinically, these are revealed by the onset of either pneumonia or pulmonary oedema (water on the lungs). Asthmatics are particularly sensitive to HCHO and their wheezing often starts around $1 \, \mu l \, l^{-1}$ HCHO.

HCHO also provides a clear example of interactions which may occur between particulate and gaseous forms of air pollution. When dust or smoke particles are also present, symptoms of HCHO are induced at even lower concentrations. This is because HCHO attaches itself to the surfaces of airborne particles which are then carried further, in concentrated form, into the lungs to induce further inflammation.

While HCHO is a confirmed cause of skin dermatitis, it is probably also a carcinogen. Largely on this basis, the TLV has been set at $1 \, \mu l \, l^{-1}$ but this also accepts that hypersensitive persons (mainly asthmatics) often experience problems well below this level.

Table 9.3   Industrial and commercial processes involving fluorine compounds which may release fluoride and HF

Emission processes	Processes using large amounts of fluorine-derivatives
Aluminium smelting	Clouding of electric bulbs
Steel production	Cut glass finishing
Phosphate fertilizers	Aviation fuel production
Enamel and pottery manufacture[a]	Insecticides and rodenticides
Brick making	Separation of uranium isotopes
Missile propulsion[a]	Synthesis of plastics
Beryllium, zirconium, tantalum and niobium purification	Aerosol, refrigerant and lubricant manufacture
Cleaning of castings[a]	Wood preservation
Welding[a]	Cement reinforcing
Sandstone and marble cleaning[a]	Furniture cane bleaching
Cryolite, fluorspar and apatite mining	Water supplementation

[a] Also involve use of HF.

Hydrogen fluoride and fluoride ions

Ubiquitous by-product

Fluorine is a gas so reactive that it does not occur naturally in elemental form. However, many fluoride-containing minerals such as fluorspar (CaF_2), cyrolite (Na_3AlF_6), and certain apatites (e.g. $Ca_{10}F_2(PO_4)_6$) are used by industry. Some industries also produce HF either as a by-product or to form various useful fluoro-derivatives (Table 9.3).

Industrial emissions are superimposed upon significant natural background sources. Consequently, levels in both air and water supplies vary widely. The majority of rural and urban air monitoring sites record very low levels of atmospheric fluoride measured as total dissolved fluoride. Near phosphate fertilizer plants, aluminium smelters, or volcanoes, however, levels may rise above 200 µg l^{-1}. Water supplies around these areas may also show elevated levels, well above the 1 µg l^{-1} recommended as an optimum to provide an 'acceptable' incidence of dental caries and at the same time allow for the correct bone growth of children.

Food and drinks are the most important sources of human fluoride intake. Normally, these contain below 1 µg l^{-1} of fluoride. Tea (3–180 µg l^{-1}), fish and other seafoods (2–85 µg l^{-1}) are heavily laden exceptions. Other vegetables and cereals grown in areas subjected to high fluoride emissions may also be enriched in fluoride.

Accumulation by plants

Crop loss in the USA due to fluoride is ranked fourth in importance after O_3, SO_2 and nitrogen-based air pollutants. However, on a weight for weight basis, fluoride is the most phytotoxic of all atmospheric pollutants. Injuries to the most susceptible plants occur at concentrations between 10 and 1000 times lower than those of other air pollutants. Rates of uptake of fluoride into leaves are also faster than those of any other pollutant and go on to cause problems to animals feeding upon these plants (see later).

Both gaseous and particulate fluorides are deposited on plant surfaces and some penetrate directly if the leaf is old or weathered. Nevertheless, the main access into a plant is by HF entering through the stomata. An important feature of fluoride uptake and transport in plants is that it is later carried in the transpiration stream towards the leaf tips or margins where it accumulates and phytotoxic effects usually develop (see Plate 5 and cover). Plant species show wide ranges of susceptibilities to fluoride but environmental factors, such as light, temperature, humidity, water stress, etc., all influence plant response. Young conifers (cover), gladioli (Plate 5), peaches (which show a defect called 'suture red spot'), and vines are especially sensitive while tea and cotton are very resistant.

There are several mechanisms which reduce fluoride levels in plants. These include shedding of individual leaves or surface waxes, leaching by rain, or volatilization. Fluoride levels are often lowest during summer months because of more favourable meteorological conditions for better dispersal of fluoride pollution and greater turnover of leaves in grass swards during summer.

There are many reports of changes in photosynthesis, respiration or metabolism of amino acids, proteins, fatty acids, lipids, and carbohydrates in plants due to fluoride (see Table 9.4 for details). Certain enzymes are modulated by the presence or absence of fluoride (e.g. enolase) but these do not explain the wide range of metabolic changes known to occur. These are due to interactions between fluoride and calcium or magnesium. Calcium and fluoride together, for example, stimulate phosphate uptake which means that calcium adsorption sites on cell membranes are involved in response to fluoride. Cytoplasmic calcium is a ubiquitous regulator of cell metabolism and many, but not all, of its effects are mediated by a calcium-binding protein calmodulin, which in turn stimulates a variety of enzymes. Moreover, calcium ions are known to affect the transport selectivity of membranes with respect to other substances. Because of this, fluoride exerts an effect on various regulatory activities (see Table 9.4) and this probably explains why it is so phytotoxic at such low concentrations.

Fluoride also forms magnesium–fluorophosphate complexes and,

Table 9.4 Physiological effects of fluoride on plants

Process	Disturbance	Likely cation interaction
Respiration and carbohydrate metabolism	Glycolysis inhibited	Mg^{2+}
	Pentose phosphate pathway enhanced	Mg^{2+}
	Unusual mitochondrial swelling	Mg^{2+}
	Oxidative phosphorylation reduced	Mg^{2+}
Photosynthesis	Unusual chloroplast structure	Mg^{2+}
	Inhibited pigment synthesis	Mg^{2+}
	Increased PEPC[a] activity	Mg^{2+}
	Reduced electron flow	Ca^{2+}
Amino acid and protein metabolism	Increased in free amino acids and asparagine	Ca^{2+}/Mg^{2+}
	Decrease in ribosome sizes	Ca^{2+}/Mg^{2+}
Nucleic acid metabolism	Changes in transcription and translation	Ca^{2+}/Mg^{2+}
Fatty acid and lipid metabolism	Increased esterase activities	Ca^{2+}/Mg^{2+}
	Decreased unsaturated/saturated ratios	Ca^{2+}/Mg^{2+}
Other metabolic changes	Increase in peroxidase activities	Ca^{2+}/Mg^{2+}
	Decrease in acid phosphatase activity	Ca^{2+}/Mg^{2+}
Transport and translocation	Altered plasma membrane ATPases	Ca^{2+}
Fruit development	Poor fertilization and seed germination	Ca^{2+}
	Reduced pollen tube growth	Ca^{2+}
	Reduced seed number and fruit size	Ca^{2+}

[a] PEPC = Phospho-enol-pyruvate carboxylase (see Chapter 8).

consequently, many enzyme pathways (Table 9.4) are adversely affected by fluoride. Most reactions involving ATP, for example, require additional magnesium complexes to function correctly. If these natural complexes are also disturbed by the presence of additional fluoride then key reactions are inhibited.

Normally, soils contain between 20 and 500 μg g^{-1} fluoride but, because it has limited solubility in soil water, uptake by roots is relatively low, and there is little relationship between soil fluoride and total plant fluoride content. Consequently, atmospheric sources of fluoride are more important than fluoride in groundwater in determining the amount of fluoride in or on a crop.

Problems associated with fluoride in plants are well known in relationship to fluorosis in farm animals. Animals grazing on pasture very close to brickworks, smelters and phosphate fertilizer factories, or fed forage gathered from such areas, may show fluorosis, a condition also occasionally found in humans (see later). The major recommendation (the Washington Standard) has been to ensure that the yearly average fluoride content of herbage does not exceed 40 μg g^{-1} a^{-1}.

Application of lime to crops and herbage has long been known to be a practical means of reducing the effects of fluoride injury. Originally, it was thought the lime caused the immobilization of the fluoride on the surfaces of the leaves as insoluble calcium fluoride (CaF_2). However, calcium chloride spraying has a similar alleviating effect to lime and recent studies have shown that the remedy actually relies upon additional calcium entering the leaves to interact with the fluoride inside and redress any calcium imbalances in the regulatory processes.

Fluorosis in animals

The USDA claims that fluorides have caused more worldwide damage to domestic animals than any other air pollutant. The Icelanders have good cause to agree with this. The 1970 eruption of the Icelandic volcano Hekla caused problems in animals within a 200 km radius and killed over 7500 sheep and lambs by fluorosis (chronic fluoride poisoning). Historical records showed that this was not an isolated event – similar problems with animals occurred after this volcano had erupted 200 and 900 years before.

Acute episodes like these are far rarer than chronic intoxication in sheep and cattle caused by eating fluoride-rich grass, hay, or silage over long periods. Inevitably, contaminated soil is also taken in as they graze. Symptoms of fluorosis in cattle follow several stages. An initial lethargic phase occurs when hides become less elastic and the animals show clear signs of pain in the rib region when pressed. This may be followed by a lameness phase with swellings and loss of milk production which leads to a rigidification of the back bones, pain in the legs and, finally, death by wasting. Pigs, sheep, and many wild animals show a similar sequence of disturbances.

Fluorosis is not confined to mammals. Silkworms are impossible to raise in the vicinity of fluoride emissions, while fluoride has replaced arsenic as the most serious cause of death to bees. Levels of fluoride in insects from fluoride-affected areas may rise as high as 400 μg g^{-1}, well above average levels of 8 μg g^{-1}, with proportional increases in death rates.

Table 9.5   Physiological effects of fluoride in animals and humans

Process	Disturbance
Carbohydrate metabolism	Glycogen levels depleted
	Glycogen turnover depressed
	Phosphorylase activity reduced
Lipid metabolism	Activation of acetate inhibited
	Liver lipases activated
	Certain esterases inhibited
Mineral metabolism	Interference in iron uptake
	Sulphite and phosphate counteract the inhibiting effect of Ca^{2+} upon intestinal absorption
Hormonal balances	Effect on parathyroid function [a]

[a] Calcium levels are influenced by parathyroid hormone (parathormone, PTH) produced by the parathyroid and a hormonal derivative of vitamin D (called 1,25-dihydroxycholecalciferol) found in the liver and kidneys – both of which raise blood serum levels of calcium. Release of calcitonin from the thyroid, however, causes enhanced calcification of the bone tissues which then reduces blood calcium levels again.

Human fluorosis, fluoridation and dental health

As already mentioned, food and drink are the main source of fluoride for the population at large, but workers in fluoride-generating industries (particularly cryolite mining) breathe fumes and dusts containing fluoride which are then absorbed into the body. The TLV is set at 3 µg l^{-1} 8 h^{-1} for HF but, because fluoride is lost in the urine, elimination rates of below 4 mg fluoride day^{-1} are used as indication of safe working conditions. Storage of fluorides takes place in the bones and teeth with normal ranges being from 50 to 500 µg g^{-1}, but these levels may rise dramatically within tea drinkers living in areas supplied with naturally fluoride-rich waters.

Understanding of the toxicity of fluoride in humans is improving. The high affinity of fluoride for magnesium, manganese, iron, calcium, and phosphate causes it to interfere with many enzymes (see Table 9.5) and affect hormonally controlled processes, especially those of the parathyroids. The variety of processes in humans affected by fluoride (Table 9.5) mirrors the range of disturbances found in plant tissues (Table 9.4). Once again, effects of fluoride are involved either with membrane-associated events involving calmodulin and calcium-dependent protein kinases or with unusual magnesium fluorophosphate complexes being formed which prevent normal magnesium and phosphate activation of enzymes.

Symptoms of human fluorosis are similar to those in animals although acute cases are very rare. Dental fluorosis is characterized by white or chalky patches on dental enamel where there has been imperfect calcifica-

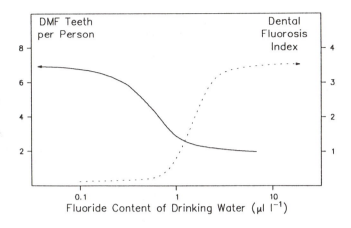

Fig. 9.2 The balance between the incidence of dental decay (where DMF = number of decayed, missing or filled teeth), due to low levels of fluoride in drinking water, and signs of dental fluorosis as a consequence of excess fluoride.

tion and a deficiency of cementing substance. In the advanced stage of chronic fluorosis in humans, this mottling may become yellow, red, brown or black. Meanwhile, tissues and ligaments may become calcified and protrusions develop on the surfaces of bones. Occasionally, these are so pronounced that they bring pressure upon spinal nerves and cause paralysis. Rigidity of the ribcage may also cause breathing difficulties. Sometimes, small pinkish-brown, bruise-like areas called *chizzola maculae* also occur on the skin during fluorosis. These disappear when sources of fluoride-contamination (air, water or food) are removed and recur when fluoride is encountered again.

It is largely problems associated with too much fluoride that have coloured arguments over fluoridation of public water supplies. Signs of fluorosis occur when fluoride exceeds 1.4–1.6 $\mu g \ l^{-1}$ in drinking water. On the other hand, the incidence of dental caries in children from areas supplied with water with fluoride lower than 0.5 $\mu g \ l^{-1}$ is between two and four times higher than the incidence in those having 1 $\mu g \ l^{-1}$ fluoride in their drinking water (see Fig. 9.2). Medical and public authorities have sought to optimize the lesser of one evil against another by advocating fluoridation of public water supplies to this level. However, the real question is one of trust. Can the water supply utilities control levels of flouride accurately enough in such large volumes and always remain within this optimal window? If so, then the public should be reassured on this point.

Organic lead

Airborne sources

A decade ago, approximately a quarter of all lead pollution was initially emitted into the atmosphere. Some of this was due to inorganic lead particulates from lead-processing industries or from the burning of certain coals which had a high lead content. However, the major air emissions then and now are in organic form as tetraethyl lead $((CH_3CH_2)_4Pb)$ or trimethyl lead $((CH_3)_3Pb)$ emitted as unburnt or partially combusted fuel vapours. The amounts involved in developed countries were huge. In the late 1960s, for example, these amounted to 180 kt of lead a^{-1} in the whole of the USA and 5 kt of lead a^{-1} in Los Angeles alone.

In most developed countries, legislation to limit the amounts of tetraethyl and trimethyl lead, used as anti-knocking agents in vehicle fuels, has been progressively introduced and newer vehicles have been redesigned to use low lead or unleaded fuels. This has significantly reduced that fraction of lead entering the environment by airborne emissions in developed countries to below one-tenth of the highest overall total in the past. However, the situation is unchanged or worsening in poorer developing countries that cannot afford the additional refining costs or newer vehicles.

Human intake

The major uptake of lead into most humans is from their food, especially vegetables grown near busy roads. Plants take up very little lead from the soil because it is usually insoluble and tightly bound to soil particles. Most lead in plants enters through the stomata in organic form. Once inside, some persists in organic form and the remainder is converted to inorganic lead. In certain circumstances, lead levels in lettuces, for example, approach 1 µg g^{-1} which means that food intakes amounting to 300 µg of lead day^{-1} may be achieved. Of this, less than one-tenth (15–30 µg day^{-1}) is taken up into the bloodstream from the digestive system.

A much higher proportion of lead (30–50%) directly enters the bloodstream by breathing. This means that people living or working near busy roads, where airborne lead levels might range from 1 to 5 ng l^{-1}, who breathe 20 m^3 of air per day are likely to absorb between 6 and 50 µg lead day^{-1} from their lungs into their bloodstreams – as much again as they might take up from their diet. Consequently, the current TLV for total airborne lead set at 0.15 ng l^{-1} is frequently exceeded and direct uptake of airborne lead often matches dietary intake, although more usually it is 20–30% of the diet in the bulk of the population not living immediately near busy roads.

The source of blood lead can be checked in some countries (e.g. the UK, Australia, and for a time, Italy) because ^{206}Pb/^{207}Pb ratios in water (1.18) and fuel (1.06) differ. Individuals living near busy roads tend to have low ^{206}Pb/^{207}Pb ratios in their blood while equivalent ratios in those living away from roads are higher.

Inorganic lead is stored in the body for some time. It accumulates in both teeth and long bones where it replaces calcium. The main route of elimination of inorganic lead involves the faeces and, to a lesser extent, the urine, sweat, hair, and nails. Organic lead, by contrast, is removed much faster in the urine.

Toxicity in the body

Symptoms due to organic lead differ from those of inorganic lead. Most effects of tetraethyl and trimethyl lead are upon the nervous system. They first show as sleeplessness and general irritability but, following heavy exposures, they progress to emotional instability and hallucinations accompanied by impaired vision and hearing. More often, at low persistent levels, they cause headaches, general fatigue and, sometimes, depression.

Inorganic lead poisoning also starts with fatigue and irritability but it is often accompanied by anaemia and abdominal pain. The anaemia is mainly due to the fact that one of the early enzymes of haem synthesis, δ-amino-laevulinate dehydratase, is specifically inhibited by inorganic lead while RBCs show increased fragility and have shortened lifespans. Chronic low-level exposure of adults to inorganic lead may increase blood pressure and, in children, disturb the metabolism of vitamin D which affects the growth of their long bones. Kidney disturbances are also associated with inorganic lead but the general slowing of nerve impulses occurs with both inorganic and organic forms of lead.

The possibility that lead may be carcinogenic remains controversial. On the weight of evidence from animal studies, inorganic lead has been placed on the list of probable carcinogens. It is also known to promote chromosomal aberrations and to affect reproduction.

Radon

Physical properties

Radon, which occurs naturally, is the heaviest known gas. It is colourless, odourless, almost inert, soluble in water (especially at low temperatures), and radioactive. It is by far the most important source of ionizing radiation to affect humans. In most developed countries, radon accounts for 40–50% of total ionizing radiation received by the population.

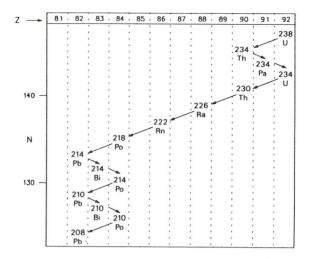

Fig. 9.3 The major decay transitions of ^{238}U to radon and then into radon daughters. α-particles are released when decays move to the left and lose two protons. β-particles are formed when movement is to the right and the number of protons increases by one.

There are 27 isotopes of radon ($^{200–226}$Rn) but only three have half-lives longer than an hour (^{210}Rn, 2.4 h; ^{211}Rn, 14.6 h and ^{222}Rn, 3.82 days). Of these, ^{222}Rn is the most important and arises from the decay of ^{238}U (Fig. 9.3). ^{222}Rn also decays into a series of radionuclides known as radon daughters or progeny. Principal among these are ^{214}Pb (half-life, 26.8 minutes), ^{210}Pb (22.3 years), ^{210}Bi (5 days) and ^{210}Po (138.4 days). The final product is ^{206}Pb which is non-radioactive. Most of the decays involved (Fig. 9.3) release α-particles but some (e.g. ^{210}Pb or ^{210}Bi) emit β-radiation.

Human uptake and hazards

The major hazards to humans of radon come from breathing in radon daughters rather than radon itself. As a gas, almost as much radon comes out during expiration as went in beforehand. Very little decays inside but, being soluble, some does enter the bloodstream by diffusion. By contrast, non-gaseous radon daughters either adhere to dust particles or form clusters around water in aerosols. Both forms are inhaled but the radon daughters are deposited along the airways where they remain and decay. Consequently, the major deposition of radon daughter-contaminated particles occurs in the pharynx and at the junctions of the bronchi (see Fig. 2.12). Their relatively short half-lives (apart from

^{210}Pb) mean that most of the radiation dose is experienced by the linings of the airways which then causes pulmonary fibrosis and cancer.

An α-particle is a doubly positive charged nucleus of ^{4}He consisting of two protons and two neutrons. It travels slower than a β-particle because of its greater mass and its energy is easily absorbed. This means that, in human tissues, α-particles are effective only over the first few millimetres. It also explains why the effects of radon daughters are confined to the epithelial layers of pulmonary tissues.

Radiation has its greatest effect on DNA. Many different types of disturbance in DNA are known to be caused by radiation but rupture of base-sugar bonds, single and double DNA strand breaks, point mutations, and chromosomal aberrations are the commonest. For some of these defects, there are natural inherent repair mechanisms (e.g. sugar-base ruptures and single strand breaks) but, over very short distances, α-particles also cause double strand breaks. In these circumstances, repair polymerases have no template on which to insert the appropriate opposite base in the other strand as in single strand repair. This means that either the damage is unrepaired or possibly the wrong bases are put into both strands. After replication during mitosis (cell division), erroneous information encoded in this particular DNA is then passed on. These errors may then appear in functional proteins or in cellular control systems affecting, for example, cell proliferation. Under certain circumstances, this may then lead to the formation of cancerous cells.

In epidemiological studies, it is sometimes difficult to distinguish the hazards associated with radon daughters from those associated with other lung cancer-inducing agents in, for example, tobacco smoke. There is also evidence that results obtained from animals experimentally exposed to radon do not accurately reflect human response. Nevertheless, estimates have been made of human lifetime lung cancer risks by using a combination of epidemiological and experimental animal studies. These are illustrated in Fig. 9.4.

Radioactivity is expressed in terms of the number of nuclei decaying per second. One becquerel (Bq) is equivalent to 1 decay s^{-1}. Different decays have different energies so damage to human tissues has to be measured in terms of combined dose. The unit used is the sievert (Sv). In the case of radon daughters, this means that a dose of 1 mSv a^{-1} is acquired by an individual exposed to 20 Bq m^{-3} over a year (i.e. the UK average). This, in turn, translates into a lifetime risk of 0.25% of contracting lung cancer as a result of radon daughters (Fig. 9.4).

Sources and countermeasures

Radon is unusual in that, being of natural origin, it is not a true air pollutant, but for convenience and effective action it is best regarded as

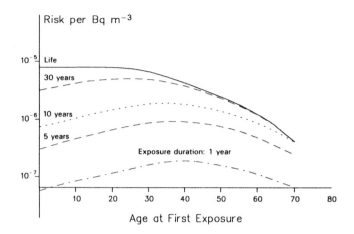

Fig. 9.4 Lifetime risk of lung cancer due to exposure to radon (from National Council on Radiation Protection and Measurement, 1984).

such. The source of radon is not homogenic except in the sense of continuing to build houses, etc., in known problem areas or for those workers who mine uranium or work in uranium-rich areas. It is now well understood that radiation arising from the decay of radon is not spread uniformly. Certain areas have much higher rates of radon release than others, but economic and commercial restraints often fail to advertise possible risks in an area until after developments have been started. Recommended action levels vary between countries (e.g. 150 Bq m^{-3}, USA; 250, Germany; 400, UK; 800, Scandinavia).

Sources of radon such as ^{238}U or ^{226}Ra are concentrated in certain acidic igneous rocks with a low melting point such as granite. It is, however, the distribution and accessibility of ^{238}U in such rocks that affect the emanation of radon. The number of faults and fissures in rocks affect the transport of radon from the solid rock into the intervening gas or liquid phases, and from there upwards – different overlying strata having different diffusional characteristics.

Radon emerges into the atmosphere by a variety of routes, including from the ground below and around the water supplies, and from natural gas or building materials. Being denser than other gases, it tends to concentrate at low points. Consequently, lower storeys of buildings in radon-prone areas have higher levels of radon than upper floors. Furthermore, most buildings have a slightly lower internal atmospheric pressure which forces radon inwards rather than outwards. This means that newer buildings which have fewer draughts are worse than older ones.

Treatment, once the problem is recognized, is relatively simple. The ground below and around can be sealed off from the rest of the building but leaks may still occur. A much better solution is to create a small porous sump area at the base of a radon-prone building and then continuously pump any gas that seeps in out of the building.

Further reading

BEIR (Committee on the Biological Effects of Ionizing Radiation) *The Effects on Populations of Exposure to Ionizing Radiation.* National Academy Press, 1980, Washington, DC.

Binder, K. and Hohenegger, M. (eds) *Fluoride Metabolism.* Verlag Wilhelm Maudrich, 1982, Vienna, Munich and Berlin.

Calabrese, E.J. and Kenyon, E.M. *Air Toxics and Risk Assessment.* Lewis, 1991, Chelsea, Michigan.

Coburn, R.F. (ed.) Biological effects of carbon monoxide. *Annals of the New York Academy of Sciences* **174**, 1970, 1–430.

Cothern, C.R. and Smith, J.E. (eds) *Environmental Radon.* Plenum Press, 1987, New York.

Filler, R. and Kobayashi, Y. (eds) *Biomedicinal Aspects of Fluorine Chemistry.* Kodansha, 1982, Tokyo, and Elsevier Biomedical Press, Amsterdam.

Murray, F. (ed.) *Fluoride Emissions – Their Monitoring and Effects on Vegetation and Ecosystems.* Academic Press, 1982, Sydney, New York and London.

National Radiological Protection Board. *Exposure to Radition Daughters in Dwellings.* HMSO, 1987, London.

Royal College of Physicians. *Fluoride, Teeth and Health.* Pitman Medical, 1976, Tunbridge Wells.

Tomatis, L. *Air Pollution and Human Cancer.* European School of Oncology Monograph (Series ed. U. Veronesi), 1990, Springer-Verlag, Berlin.

Waldbott, G.L. *Health Effects of Environmental Pollution* (2nd edn). C.V. Mosby, 1978, St Louis, Missouri.

Weinstein, L.M. Fluoride and plant life. *Journal of Occupational Medicine* **19**, 1977, 49–78.

World Health Organization. *Fluorides and Human Health.* WHO, 1970, Geneva.

Chapter 10

RECENT FOREST DECLINE

'In nature there are neither rewards nor punishments – there are consequences.'
in 'Some Reasons Why' (Chapter vii, Lectures & Essays, 3rd Series) by Robert G. Ingersoll (1833–1899).

Occurrence and classification

Over the last three decades, extensive damage to forest trees has occurred over wide areas of Europe and North America. Sometimes, the symptoms can be explained by well-known agents which damage trees (e.g. insect pests, fungal attack, deficiencies of magnesium, etc.). In many areas, however, the causes of premature leaf or needle loss and tree death are far from clear. This unexplained phenomenon has been given several different names. The most cautious is 'recent forest decline' derived from the German phrase *'neuartige Waldschäden'* meaning new kind of forest damage. The more emotive *'Waldsterben'* (forest death) or tree dieback are now less common.

Recent forest decline first appeared in Germany in the early 1970s, initially on silver fir (*Abies alba*) and then on Norway spruce (*Picea abies*). More recent effects on pines (*Pinus*), beech (*Fagus*), chestnuts (*Castanea*), and oak (*Quercus*) are probably due to similar causes. The current extent of the problem is illustrated by survey data relating to many CEC countries (Table 10.1). Many species are still in decline but there is some improvement in some species (e.g. *Quercus ilex*). However, it may be significant that recent forest decline is more evident on west-facing slopes facing the prevailing wind in central Europe rather than on those facing east where higher levels of SO_2 and nitrogen oxides are emitted. Similar types of forest decline also occur in north-east USA, especially with red spruce (*Picea rubens*), and in the eastern provinces of Canada with sugar maples.

Until the existence of the problem was widely accepted, surveys from forest to forest, from area to area, and from country to country were not correlated because different methods of assessment of the degree of damage were used. However, a standard classification of the damage on a scale ranging from 0 (Healthy) to 4 (Dead) is now used by CEC countries. The essential details of this classification system are listed in Table 10.2 although considerable variations exist between tree

Table 10.1 Recent % defoliation of different tree species in Europe.[a] Data include surveys of 67 335 trees over 4 years (CEC Forest Health Report, 1991)

Tree species	Defoliation[b]							
	0–10%				> 25%			
Year:	87	88	89	90	87	88	89	90
Castanea sativa	73	78	67	59	6	7	9	19
Fagus sylvatica (Beech)	63	67	64	55	13	9	11	18
Picea abies (Norway spruce)	64	62	60	56	24	26	27	30
Picea sitchensis (Sitka spruce)[c]	48	39	28	19	21	24	14	17
Pinus halepensis (Aleppo pine)	61	54	65	64	11	8	6	5
Pinus nigra (Black pine)	73	71	78	61	6	5	3	11
Pinus pinaster	66	67	68	70	14	12	10	19
Pinus sylvestris (Scots pine)	67	58	58	53	9	10	10	13
Quercus ilex	54	59	64	71	18	7	5	5
Quercus petraea (Sessile oak)	69	70	65	61	6	9	10	9
Quercus pubescens	88	78	71	62	8	7	8	16
Quercus robur (English oak)	49	41	53	61	22	24	14	17

[a] All CEC (including former East Germany) plus Austria, Czechoslovakia, Hungary and Switzerland.
[b] 11–25% may be obtained by difference.
[c] Loss of needles mainly due to severe attacks by green spruce aphid (*Elatobium abietinum*).

species because their shape, growth characteristics, and symptoms differ.

If Norway spruce is taken as an illustrative example, there is a great danger of being highly selective over the choice of pictures and thereby over-dramatizing the problem. The typical example shown in Plate 10 illustrates a damaged (Class 3) Norway spruce tree showing progressive needle loss and a distinctive streamer-like hanging of the secondary branches which becomes less obvious once a large number of needles are lost.

Recent forest decline is also characterized by a yellowing of the older needles (Plate 11) which occurs especially on trees at higher elevations in the granitic central regions of Europe. One marked feature of this

Table 10.2 System of classification of trees according to health as recommended by the CEC

Class	Needle (or leaf) loss (%)	Description
0	0–10 [a]	Healthy
1	11–25 [b]	Slightly damaged
2	26–60 [b]	Medium to seriously damaged
3	61–99	Dying
4	100	Dead

[a] Where existing needles (or leaves) between 26–60% are yellowed then one class higher, above this two classes higher.
[b] Where over 25% of existing needles (or leaves) are yellowed, one class higher.
N.B. Optional parameters which may be included during classification include crown deformation, trunk condition, annual terminal shoot length, age of oldest needles, fall of healthy twigs, extra twigs, etc.

yellowing is that the upper surfaces of the needles are yellower than underneath – as though the 'problem' were coming from above. In fact, this particular type of injury is not a recent phenomenon and occurred in the Adirondacks of the USA back in the 1950s and in the Black Forest of Germany in the early 1960s. What is different now is that these yellowing symptoms occur over much wider areas and in areas which are not naturally deficient in magnesium. Similar symptoms are well known in magnesium-poor areas and may be traced directly to magnesium deficiency in the trees, which is aggravated by acidic deposition. Normally, treatment of forest soils with magnesium-based fertilizers or dolomitic limestone corrects this problem in these particular areas.

This also illustrates one of the problems with recent forest decline and the classifications of damage. It is far from easy to distinguish an 'unknown' cause of tree injury from that caused by known stresses (e.g. mineral deficiency or pathogens). Table 10.3 lists a number of the commoner agents that affect Norway spruce which can be superficially confused with recent forest decline and which have led to controversy among experienced foresters. Indeed, some still ask the question 'Is recent forest decline a distinctly different phenomenon?'. On balance, when the surveys have been done carefully and all the effects of other agents have been eliminated (including magnesium deficiency), the answer is 'Yes, but the exact cause(s) are still unknown'.

Dendrochronology is the technique of using of tree ring data to provide information on age and past performance of trees. It has been used, for example, to follow the progress of problems associated with air pollution. One of the best, from which the results shown in Fig. 10.1 are taken, was done in the Rhône valley where there are sources of both SO_2 and fluoride. By combining observations of the width of tree rings with a computer analysis, which takes into account monthly rainfall and temperature, the start of the damage in this instance could be traced

Table 10.3 Known causes of damage to Norway spruce (*Picea abies*) which can be confused with recent forest decline damage

Agent	Symptom
Abiotic:	
Magnesium deficiency	Older needles yellow or chlorotic on upper surfaces
Low temperatures	Needles of all ages slightly chlorotic and become yellow in early spring
Drought or salting of nearby roads	Red coloration of older needles on branches in exposed localities
Pathogenic:	
Spruce needle cast fungus (*Lophodermium piceae*)	Older needles red-brown with black spots or bands in spring
Spruce needle scorch (*Lophodermium macrosporum*)	Black spots along central rib of needle
Rhizosphaera needle scorch (*Rhizosphaera kalkhoffii*)	Like the above symptoms but with much smaller black spots
Spruce needle rust (*Chrysomyxa abietis*)	Orange banding of needles
Grey mould (*Botrytis cinerea*)	New needles go brown and hang down. Isolated patches of infection. Not to be confused with frost damage
Sawfly (*Pristiphora abietina*)	Young needles eaten on one side only, rest turns red
Spruce bell moth (*Epinotia tedella*)	Leaf bases eaten and needles turn red-brown
Spruce bark beetle (*Ips typographus*)	Bark peeling and resin droplets on trunk, trees ultimately become red-brown and die

back to the time when another aluminium smelter came into operation in the area.

Dendrochronological studies of recent forest decline have revealed patterns of reduced tree ring growth dating back over 40 years. They confirm that fluctuations of climate were not responsible. They do, however, show that most reductions in growth occurred first at higher altitudes and gradually spread down the slopes of afforested regions. Tree ring studies, however, are ambivalent as to cause except in certain cases (e.g. Fig. 10.1). This is because it is impossible with this technique to distinguish clearly one specific factor (e.g. magnesium deficiency) which could be responsible for the reductions in tree growth.

Possible causes

Much effort has been devoted to the identification of possible causes of recent forest decline. Table 10.4 lists the major hypotheses of cause roughly in the order in which they have been proposed.

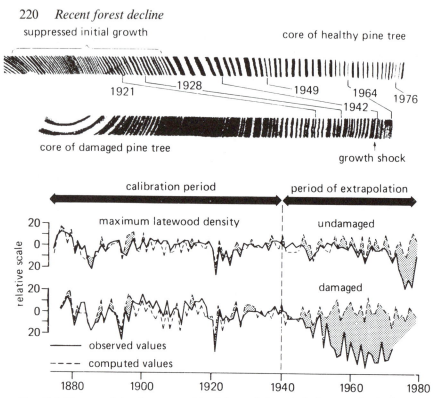

Fig. 10.1 Cores and densitograms of healthy and damaged pine trees in the lower Rhône Valley, France. In this particular case, the reduction of growth in the damaged cores is directly related to the introduction of an aluminium smelter to the vicinity around 1947 which emitted SO_2 and fluorides (courtesy of Dr H. Flühler, Birmensdorf, Switzerland and D. Reidel Publ. Co., Dordrecht, The Netherlands).

The bad-practice hypothesis

It is a long-held view of certain foresters that there were many inherent dangers in establishing large forests (or replanting old ones) with a single species, often of the same provenance. Such monocultures are held to be ultimately more sensitive to attack by natural pathogens, especially as trees reach maturity.

These 'I-told-you-so' predictions are not very helpful, 60 years or so later, as an answer to the cause of recent forest decline. Perhaps their use lies in any fresh consideration of future planting strategies. On the other hand, it may simply be that some forest trees have been allowed to grow too old before felling and replacement. Many of the affected trees are now well over 60 years old and more prone to stress anyway.

Table 10.4 Summary of hypotheses suggested as possible causes of recent forest decline (much abbreviated, see text for details)

Hypothesis	General features
1. Bad practice	Poor forestry practice including use of foreign species on unsuitable soils
2. Acid rain/soil leaching	Acidic (mainly wet) deposition (excluding simple magnesium deficiency) causes soil leaching of potassium, etc., and increases the exposure of roots to aluminium ions
3. Ozone/photochemical	Action of sunlight causes enhanced production of ozone, PANs and hydrogen peroxide, etc., which promote attack on cell membranes
4. Modifier or pollution-enhanced stress/infection	A series of droughts or unusually cold winters induce persistent attacks by pathogens (including known viruses), along with general atmospheric pollution; weakens still further aged, sensitive and exposed trees
5. Ammonium/excess nitrogen	Over-saturation of trees with ammonium ions, etc.; they then grow too fast and are thus attacked by parasites or become more prone to stresses
6. Chloroethene/photoactivation	Presence of certain CFCs initiates UV light-initiated damage in cell membranes
7. Alternatives (many)	Wide variety of possibilities including tetraalkyl lead from car exhausts exerting toxicity when degraded to trialkyl lead, phytotoxic dinitrophenols, PCBs and aromatics, latent unidentified viral infections
8. Stress–ethene/multi-interaction	Any of the above causes linked together by a common response such as the formation of stress–ethene which then reacts with O_3 to form ozonolytic HHPs which disrupt cell membranes

The acid rain or soil leaching hypothesis

This possible explanation, largely attributed to Professor Ulrich of Göttingen, suggests that, when excessive acidity reaches forests, detrimental changes in soil chemistry occur which affect root growth and, ultimately, threaten the health of the whole tree. Initially, extra inputs of sulphur and nitrogen increase tree growth but, in the longer term, acid deposition reduces the soil buffering capacity (see Chapter 5). The ability of the soil to continue to provide essential nutrients such as calcium or potassium for tree growth is then diminished as these are leached from the soil around roots and lost in run-off. Losses of magnesium (and manganese), which cause yellowing of leaves or needles in granitic areas known to be low in magnesium, etc. (see above), are also accelerated by acidic deposition. The hypothesis goes on to suggest that aluminium ions are then mobilized from soil particles into solution as the pH changes (as discussed in Chapter 5). These aluminium ions are toxic both to uptake mechanisms on the fine root hairs and to the fungi in mycorrhizal associations around the roots. This, in turn, inhibits the ability of roots to take up essential minerals or water and trees become more susceptible to disease or drought. Increased needle loss, decay, etc., then trigger off other acid-forming mechanisms which make the problem worse.

There is considerable evidence to support this hypothesis. Many studies have shown higher levels of H^+ and aluminium ions as well as reduced levels of calcium and potassium (plus magnesium and manganese) when simulated acid rain has been applied to forest soils. This hypothesis is consistent with a known cause of forest decline whereby granitic forest soils become more magnesium-deficient with acidic deposition. As a result, these trees have greater numbers of yellow second-year needles which have abnormally low magnesium contents even though the original soil had enough magnesium.

At the moment, if the acid deposition effects on known magnesium-deficient forest soil areas are excluded, this hypothesis in its original form remains unproven because a number of studies have been unable to find a clear relationship between acid deposition, aluminium mobilization, and recent forest decline. It could, however, be a partial explanation when combined with some of the other possibilities (see later).

The ozone or photochemical hypothesis

This hypothesis suggests that the needle (or leaf) damage associated with recent forest decline is initiated by higher levels of O_3 (Chapter 6) which do not decay away as fast at higher altitudes as they do close to sea level (Fig. 6.3). Increased levels of H_2O_2 and other secondary photooxidants derived from O_3 have also been suggested as possible causes of needle (and leaf) damage. Distinct effects of O_3 and other oxidants upon vegetation have already been covered (Chapter 6) but this recent forest decline hypothesis goes on to suggest that damaged cell membranes then allow leaching of nutrients from the needles or leaves which encourages subsequent attack by insects and fungi.

On first reflection, direct attack of photochemical agents might explain the enhanced yellowing of needles sometimes associated with recent forest decline and especially the 'damage-uppermost' symptom often observed in the secondary branches. However, experimental fumigations of spruce with O_3 for single seasons have failed to produce visible symptoms associated with recent forest decline, although decreased rates of photosynthesis in fir have been detected. Scots pine (*Pinus sylvestris*), however, is O_3-sensitive and a single season of fumigation causes chlorotic mottling of the needles as well as reduced root growth and increased senescence of older needles.

Longer fumigations of spruce over several seasons have been more informative. Five-year-old Sitka spruce fumigated for three summers with O_3 show classical symptoms of recent forest decline (i.e. yellowing of two-year-old needles) at the start of the fourth year of fumigation (Plate 12). One aspect of overwintering is the ability of trees to harden and resist frost injury effectively. It is now established that previous summer O_3 fumigations reduce the ability of trees to harden correctly. Trees that do not harden to sufficiently low temperatures then have their cells ruptured by ice crystal formation and, when spring arrives, problems of leaching arise. Perhaps changes like these which involve both a pollutant (e.g. O_3) and a natural stress (low temperatures) are better covered by the modifier hypothesis (see below).

The ozone (or photochemical) hypothesis is a plausible explanation but better experimentation is still required to verify or refute it as a primary cause by itself. It is still likely that the relative importance of O_3 as a single source of environmental stress varies between different areas and species.

The modifier or pollution-enhanced stress/infection hypothesis

Sulphur dioxide in gaseous form has been discounted as a primary cause of recent forest decline. In fact, levels of SO_2 in affected areas have declined recently. Needle-depleted branches of affected trees are often covered with lichens, although the particular species of lichens involved are those more tolerant of acidic conditions. Similarly, there is little evidence that nitrogen oxides alone could cause recent forest decline although they may contribute excess nitrogen (see next section). Moreover, climatic fluctuations over the last century are not sufficient in themselves to explain the increased incidence of recent forest decline. This is despite the fact that symptoms of recent forest decline have probably been exacerbated by a recent series of dry summers.

The case for interactions between SO_2, nitrogen oxides, and O_3 is much better. As Chapter 11 later shows, such mixtures cause more-than-additive reductions in growth and faster appearance of injury in a variety of plants. Indeed, mixed exposures could cause the needles or leaves to be more readily attacked by pathogens or predisposed to stress injury.

In the regions affected by recent forest decline, some of the most important stresses (e.g. frost injury or winter desiccation, drought, and wind) are interrelated. For example, wind increases transpiration, which is undesirable during drought, and low temperatures render water unavailable for use because it is frozen. It is known that rates of water uptake by roots are greater in those plants that are subjected to atmospheric pollution and that this is made worse because mixtures of pollutants affect root/shoot ratios. In other words, they decrease the relative proportion of roots and increase the relative amounts of the affected plant above ground. This means that a reduced amount of root tissue works much harder to sustain a bigger proportion of above-ground tissue during droughts. Consequently, trees are more susceptible to water stress which is a very real possibility in forest soils that do not retain large amounts of water.

To summarize, the modifier hypothesis basically asks the question 'Does the adverse cold, drought or wind stress (or sustained infection) predispose the trees to damage by the mixture of atmospheric pollutants or has their additional presence rendered them more susceptible to subsequent stresses or infections?'. The answer is still not known. However, it is more than likely that it does not matter which comes first – one predisposes trees to the other.

The ammonium or excess nitrogen deposition hypothesis

One of the most evident changes in the patterns of air pollution emission has been a shift in the balance from SO_2 towards N-based air pollutants as a result of increased (mainly mobile) combustion (Chapter 3), increased applications of artificial fertilizers, more sewage treatment, and intensive animal husbandry (Chapter 4). High levels of atmospheric NH_3 are now evident in the Netherlands and surrounding countries (Chapter 4) which increases wet and dry deposition of N onto sensitive ecosystems including forests.

With this in mind, Professor Nihlgård of Lund has put forward a wide-ranging hypothesis to account for recent forest decline. This postulates that, due to the extra N, there is an initial burst of growth which then leads to an imbalance between carbohydrates and proteins. Valuable minerals (phosphate, magnesium and potassium) are then consumed as the plant attempts to re-establish a balance while toxic by-products (amides, amines and ammonium derivatives) increase cell leaching and encourage fungal attack or growth of epiphytic lichens. These, in turn, reduce photosynthesis.

At the same time, roots are deprived of carbohydrates, because these are needed above to compensate for the extra N, which means less carbohydrates are available for mycorrhizal associations of roots with specific soil fungi. This reduces still further the amount of microbial activity in the soil which is already depressed because the extra N coming into the soil reduces N_2 fixation. Such changes caused by reduced root and enhanced leaf growth then render forest trees more susceptible to other stresses imposed by drought or wind-blow, or to infection by pathogens.

A supplementary explanation based on this initial hypothesis has also been made by Professor Schulze of Bayreuth. He is convinced that the major form of N taken up by conifers is NH_4^+ rather than nitrate. Forest soils are naturally short of both but wet deposition of both nitrate and NH_4^+ replenishes these levels. Nitrate is lost in run-off but NH_4^+ ions are taken up by roots and compete with magnesium ions. Consequently, trees growing on soils with adequate levels of magnesium become seemingly magnesium deficient due to uptake of extra NH_4^+ ions rather than by magnesium being in short supply. Moreover, uptake of NH_4^+, but not nitrate, also takes place through needle surfaces which also causes magnesium leaching, but the net effect is the same in needles as in roots – symptoms of magnesium deficiency.

There are strong indications to support this hypothesis and this magnesium–ammonium competition codicil. N-based atmospheric pollutants contribute around 30% of acidic wet precipitation. After heavy snowfall and snowmelt, NH_4^+ levels are raised and this often coincides with the most sensitive stages of plant development. It also explains why soils containing magnesium can grow trees which show symptoms of magnesium deficiency.

The chloroethene or photoactivation hypothesis

The novel nature of recent forest decline has led Professor Frank of Tübingen to suggest that freshly introduced atmospheric pollutants are responsible. Such new chemicals must have half-lives in the troposphere sufficiently long to be transported to rural areas at higher altitudes and to be toxic to plants at relatively low concentrations. His research team examined a wide range of CFCs (see Chapter 7) and concluded that the chloroethenes ($CHClCCl_3$ and CCl_2CCl_2) can induce phytotoxic effects similar to those associated with recent forest decline. This damage is especially severe in the presence of strong light enriched in UV around 280–320 nm. This type of radiation mainly occurs at high altitudes but is strongly attenuated by misty, lower urban and semi-urban atmospheres.

According to this hypothesis, the toxic effects of chloroethenes first start in the lipid layers of plant cells, where they naturally accumulate, and UV light converts them into reactive chlorine atoms, dichloroacetylene, and even phosgene which then produce the characteristic yellowing often associated with recent forest decline.

There have still been no comprehensive studies to evaluate chloroethene-induced injury as a primary cause of recent forest decline. Some authorities dismiss this possibility because levels are far too low but miss the point that CFCs are extremely persistent because there are no organisms to degrade them. Furthermore, they are concentrated by a number of processes. Calculations show, for example, that tetrachloroethene in gaseous form around a leaf may attain a 2000-fold increase in concentration in the wax coatings of leaves. This particular hypothesis warrants more attention than it has hitherto attracted.

Alternative hypotheses

There are many other hypotheses to explain recent forest decline – well over 100. There are many like the chloroethene hypothesis which claim that organic micropollutants of various types might be the cause. From the 800 or so organic compounds that have been identified in air samples, there are a number of homogenic emissions which could be harmful. For example, there are occasional rainfall episodes when the levels of trialkyl lead are very high (see Chapter 9) and, at high levels, they could be phytotoxic.

Other claims have been made for the dinitrophenols, which are potent inhibitors of respiration and mineral uptake, and for a variety of other groups of organics such as the polyaromatic hydrocarbons (naphthalene, anthracene, benzpyrene, etc.) which are partially oxidized before exerting their phytotoxic effects. Similar hazards to forest trees have been attributed to the persistence of pesticides and herbicides (e.g. polychlorinated biphenyls or PCBs).

In complete contrast is the possibility that some unidentified biological infection may be responsible for initiating recent forest decline. Bacterial and fungal attack were very quickly eliminated as initiators of recent forest decline. However, the possibility of a less well-known virus and its vector propagating a disease throughout forest trees has remained. This is difficult to disprove because the virus particles could have been spread much earlier and resided in a dormant phase for some time before conditions were ripe for active phases of the 'disease'. There are many examples of widely spread, and apparently innocuous, viral particles throughout the plant kingdom. The *Potex* virus, for example, has been found in reasonable proportions (2–22%) in all those tree species known to be affected by recent forest decline. How any viral infection would spread is not known but it could be by insects, needle spores, root mycorrhizal hyphae, or animals in the soil.

At present, the nature and spread of recent forest decline does not fit well with most of the characteristics shown by known viral infections of plants, and this hypothesis is not viewed as a likely possibility.

The stress–ethene interaction hypothesis

It is more than likely that recent forest decline is another facet of problems like climate change (Chapter 8) and O_3 holes (Chapter 7) that arise from the activities of modern society which become more complex and increasingly more threatening to the natural environment. Probably, the answer to recent forest decline will never come in a clear-cut manner favouring one hypothesis over the others. Some degree of interaction between various possible causes is most likely. Indeed, several features of the soil-leaching, extra nitrogen, modifier and photochemical hypotheses overlap each other.

One avenue for progress is to seek common mechanisms which link these features. One of these may come through study of the different ways plants respond to stress. These often involve ethene (C_2H_4 or ethylene), a natural plant growth substance which, in trace amounts, may interact with other regulators to coordinate a variety of developmental responses. Experimentation has shown, for example, that short O_3 fumigations (150 nl l^{-1} for 7 h) cause visible leaf injury to clean-air grown seedlings the day after treatment. If, however, similar seedlings are grown all the time in O_3 atmospheres they are quite undamaged. This tolerance is related to the fact that these preconditioned seedlings only produce low amounts of ethene whereas those injured by a burst of O_3 produce large quantities of 'stress' ethene. Emissions of extra ethene can be prevented by treating the plants with an inhibitor (amino-ethoxyvinylglycine) before the O_3 fumigation which, in turn, protects plants from subsequent O_3 injury.

Sudden exposure of similar seedlings to nitrogen oxides (instead of O_3) also causes the production of similar amounts of 'stress' ethene to that caused by a sudden exposure to O_3 but this does not cause injury. This probably means that O_3 and ethene react to generate HHPs (see Chapter 6) which then causes ozonolysis, leaching, and visible leaf injury.

Ethene emissions are enhanced by different environmental stresses as well as the presence of other air pollutants. Many of these are involved in hypotheses already discussed. For example, mineral deficiencies (soil leaching hypothesis), chilling, wind and drought (modifier hypothesis) all increase rates of ethene evolution. These could then react with O_3 to release HHPs which would enhance ozonolysis and leaf injury. This unifying explanation therefore combines aspects of several of the hypotheses already mentioned and may be part of the solution to this complex problem.

Further reading

Bauer, F. (ed.) *Diagnosis and Classification of New Types of Damage Affecting Forests.* Special Edition, Commission of the European Communities, DGVI, F3, 1984, Brussels.

Frank, H. Airborne chlorocarbons, photooxidants and forest decline. *Ambio* **20**, 1991, 13–18.

Johnson, D.W. *et al.* Soil changes in forest ecosystems: evidence for and probable causes. *Proc. Royal Soc. Edinburgh* **97B**, 1991, 81–116.

Nihlgård, B. The ammonium hypothesis – an additional explanation to the forest dieback in Europe. *Ambio* **14**, 1985, 2–8.

Prinz, B. Causes of forest damage in Europe. *Environment* **29**, 1987, 11–37.

Rehfuss, K.E. Perceptions of forest diseases in central Europe. *Forestry* **60**, 1987, 1–11.

Schulze, E.-D. and Freer-Smith, P.H. An evaluation of forest decline based on field observations focussed on Norway spruce, *Picea abies. Proc. Royal Soc. Edinburgh* **97B**, 1991, 155–168.

Smith, W.H. *Air Pollution and Forests. Interactions between Air Contaminants and Forest Ecosystems.* Springer-Verlag, 1981, New York, Heidelberg and Berlin.

Ulrich, B. and Pankrath, J. (eds) *Effects of Accumulation of Air Pollutants in Forest Ecosystems.* D. Reidel, 1983, Dordrecht, The Netherlands.

Chapter 11

INTERACTION AND INTEGRATION

'Est modus in rebus, sunt certi denique fines,
Quos ultra citraque nequit consistere rectum.

'Things have their due measure; there are ultimately fixed limits,
beyond which, or short of which, something must be wrong'
from 'Satires' (Book I, v. 106) by Horace (65–8 BC)

Interactions

Uncertain contributions of stress

Earlier chapters have identified individual pollutants that are causes of
specific problems in plants or animals, or with materials. For example,
adverse changes in water quality brought about by acidic precipitation
have been shown to be indirectly responsible for the precipitous fall in
freshwater fish stocks in certain sensitive areas. Nevertheless, there are a
number of environmental problems like recent forest decline that cannot
be attributed directly to a single gaseous pollutant, or to acidic precipitation, or even just to air pollution. Very often, two or more factors
combine together to bring about adverse changes.

Any environmental factor that is unfavourable to living organisms is
called stress and the stress resistance of an organism is defined as the
ability to survive stress. Such stresses are often subdivided into biotic
and abiotic forms to distinguish those caused by other organisms from
those of a physical or chemical nature. Table 11.1 lists important abiotic
and biotic stresses in plants, but many of these abiotic stresses also affect
animals although their biotic stresses are very different.

Synergistic or more-than-additive?

With so many different types of stresses, of which some coincide and
others occur in sequence, the outcome is less predictable than with a
single stress. In some cases, two stresses may have different modes of
action which have no effect upon each other and go on causing the
appropriate amount of change which when added together is equal to
the sum of the two separately. Such an effect is termed 'additive' and is
not an interaction. If, however, the sum of the two effects is less than

Table 11.1   Different types of plant stress

Stress	Type	Variations
Abiotic	Temperature	Low (frost and winter cold or chilling); high (heat)[a]
	Water (and humidity)	Deficit (drought); excess (flooding)
	Radiation (intensity, wavelength)	Infrared; visible and UV (inc. UV-B and O_3 holes); ionizing (radioactivity)
	Chemical	Salt (sea-spray, road salting, etc.); SO_2, NO, NO_2, O_3, PANs, NH_3, H_2S, CO_2[a], etc.; acidic wet deposition; particulates; herbicides and insecticides
	Others	Seasonal, wind, pressure, sound, magnetic, electrical
Biotic[b]	Animalia	Insects, grazing herbivores or nematodes
	Fungi	e.g. rusts, mildew, etc.
	Higher plants	e.g. mistletoe
	Protista (unicellular eukaryotes)	e.g. yeasts, protozoans
	Monera (prokaryotes)	e.g. bacteria
	Viruses	e.g. TMV
Edaphic[c]	Trace elements	N, K, phosphate, etc.
	Heavy metals	Lead, cadmium, copper
	Mycorrhizal associations	
	Compaction	
	Weathering	

[a] Includes global warming implications.
[b] Assumes the five-kingdom classification system of Whittaker.
[c] Stresses associated with soils which may also include some of the abiotic and biotic stresses listed above.

those caused by the two individual stresses separately, the interaction is known as 'antagonistic'. In other words, one stress apparently alleviates the effect of the other. Alternatively, the sum of the combined effects may be greater than the two separate effects added together. In this case, the interaction is sometimes called 'synergistic'. The exact definition and the proof that 'synergism' has taken place is controversial because the word can so easily be misused. The use of the term 'more-than-additive' instead is probably better in most instances.

The real problem lies in defining and determining how relative amounts of stress are measured. In the case of air pollution, each pollutant can be measured in terms of concentration ($nl\ l^{-1}$) or of dosage ($nl\ l^{-1}\ h^{-1}$) and this type of interaction is best illustrated by a hypothetical experiment. Suppose a group of plants (S) is exposed for 72 h to a concentration of

50 nl l^{-1} of SO$_2$, another group (N) to 50 nl l^{-1} of NO$_2$, and yet another (S + N) to a mixture of 50 nl l^{-1} of SO$_2$ plus 50 nl l^{-1} of NO$_2$ simultaneously. At the end of the fumigation, growth of all three polluted groups (S, N, S + N) is then compared to growth of a fourth group grown in clean air (CA). The depression of growth in group S turns out to be 20%, that of group N to be 10%, and that of group S + N 60% relative to CA. A suitable statistical test would then probably show (if enough replicates were undertaken) that an interaction has taken place which some would claim is synergistic.

The main difficulty with this experiment is deciding if the treatment $S + N$ is the correct mixed treatment for this experiment. The concentration (or dosage) has actually doubled and may have passed a threshold which changes plant response. Perhaps 25 nl l^{-1} SO$_2$ plus 25 nl l^{-1} NO$_2$ ($S/2 + N/2$) for 72 h is the equivalent dosage? In order to be absolutely sure, two mixed treatments ($S + N$ and $S/2 + N/2$), plus S and N, and two controls, clean-air (CA) and charcoal-filtered (CF), are required. If all four tests ($S + N$, N, S and CA; $S + N$, N, S and CF; $S/2 + N/2$, N, S and CA; $S/2 + N/2$, N, S and CF) show significant interactions then synergism between two pollutants exists.

In practice, however, such definitive experimentation can rarely be achieved or justified and, consequently, only the use of the term 'more-than-additive' in this instance is appropriate. However, this approach only applies to comparisons between 'similar' types of stress (in this case two or more air pollutants) because both the chemical natures, and ultimate dosages of pollutants are involved. In the study of an interaction between, for example, SO$_2$ and drought stress, the concentration (or dosage) of the pollutant remains the same in the single and mixed treatments. Consequently, a distinction between more-than-additive or synergistic is not necessary.

The experimentation above also illustrates another point made much earlier in Chapter 2 about the choice of units to express concentrations of atmospheric pollutants. By using nl l^{-1} (volume per unit volume), the direct comparability of SO$_2$ to NO$_2$ on a molecule-to-molecule basis at any temperature and pressure can be appreciated. If, however, each concentration is expressed as mass per unit volume (e.g. 131 µg m^{-3} SO$_2$ and 94 µg m^{-3} NO$_2$ only at 25°C and 101.3 kPa in the above experiment) the equivalence is not immediately clear.

Effects of mixed pollutants on plants

Up to the 1980s, research interest in Europe concerning the effects of ambient air pollutants on plants concentrated heavily upon SO$_2$ while in North America rather more interest was paid to the effects of O$_3$. This is because these pollutants were then recognized as being a major part of the pollution problems in these particular areas.

More recently, Europeans have realized that nitrogen oxides now accompany SO_2 at greater concentrations and, under suitable meteorological conditions, conditions for O_3 formation may exist anywhere in Europe. Ironically, the success with which legislation has reduced the levels of particulates and smoke from the atmosphere over Europe has now extended those areas where O_3 is more likely to occur.

In North America, the co-occurrence of SO_2 and nitrogen oxides emissions is now recognized, although it is still believed that O_3 is generated some distance from sources of nitrogen oxides and that combinations of all three are less likely than in Europe. In Chapter 6, the daily pattern of photochemical smog is described in some detail and a typical example, characteristic of this daily phenomenon, is illustrated in Figure 6.3. The main feature of photochemical smog is that little SO_2 is emitted and it requires several non-windy, warm, bright days in succession to be established. In many industrial parts of the world, however, these conditions are infrequent and patterns shown in Fig. 1.4 occur, especially in Europe. In these mixed conditions, levels of photochemical oxidants are usually lower but emissions of SO_2 and nitrogen oxides for power generation occur immediately after transport of people to work, which involves additional nitrogen oxide emissions.

Uptake of pollutants into plants takes place mainly through stomata (see Chapters 2, 3 and 6). This means that, in urban and semi-urban environments, the atmospheres most likely to affect plants are those prevailing between 8 am and 4 pm, especially those around midday when light intensities and gaseous fluxes are greatest. Over longer timescales, presence of leaves on deciduous plants and their period of most active growth (late spring) is when they are most sensitive to pollutants. For evergreens (conifers, grasses, etc.) this period of sensitivity may be longer.

Most studies of interactive effects involve just two pollutants (SO_2 + O_3, SO_2 + NO_2, NO_2 + O_3 or NO_2 with NO). Studies involving three or more pollutants simultaneously are very rare. The literature describing such studies of interactions reveals a wide and bewildering array of additive, more-than-additive, or antagonistic effects. This is to be expected because each study has a choice of different pollutants, concentrations, and durations of exposure. Furthermore, fumigations have been done using various species, cultivars or clones and often measure quite different parameters of change (i.e. dry or fresh weight gain or loss, grain yield, leaf injury, net photosynthesis, enzymatic change, etc.).

Rather than bury the salient observations in a mass of detail, Table 11.2 summarizes the most important interactions in plants that frequently occur in the presence of mixed pollutants. Once again, these should be treated only as a guide. Exceptions always exist and the various permutations of treatment and species, etc., always introduce uncertainty. Such

Table 11.2   Known interactions caused by atmospheric pollutant mixtures

Mixture	Interaction	Characteristic affected
$SO_2 + O_3$	More than additive	Leaf appearance (grapes, radish, peas, tobacco, pines)
		Root growth (tobacco)
		Yield (soya, alfalfa, poplar)
		Stomatal conductance (soya)
	Less than additive	Fungal infection
		Leaf appearance (apple, soya, tomato)
		Yield (alfalfa, rape)
		Needle length (pines)
$SO_2 + NO_2$	More than additive	Pollen tube growth (grasses)
		Yield (grasses, barley)
		Enzyme activity (peas)
	Less than additive	Yield (rape)
$NO_2 + O_3$	More than additive	Pollen tube growth
		Translocation to roots reduced (beans)
	Less than additive	Yield
$SO_2 + NO_2 + O_3$	More than additive	Premature leaf fall (poplar)
		Leaf appearance (radish, peas)
		Leaf area (rape)
$SO_2 + HF$	More than additive	Leaf appearance (barley, maize)
$O_3 + HF$	More than additive	Leaf appearance (*Coleus*, mint)
$O_3 + H_2S$	Less than additive	Photosynthesis (barley, maize, citrus)

studies of interactions always add to knowledge and some of this is very useful, but unless these investigations are accompanied by attempts to understand the underlying mechanisms no real progress is ever made.

An alternative approach is to attempt to understand the variations in the chemistry, biochemistry and physiology associated with the different single pollutants as has been outlined in previous chapters. Another is to develop a model to assist understanding which, if it is sufficiently robust, will be adaptable and predictive. Such ideas are developed later in this chapter. The third approach is best called the 'inspired guess' approach in which the few clues showing that responses to one pollutant (or stress) may differ in the presence of another are examined in more detail (see next section). This approach has had some success.

Mechanisms of pollutant interaction

One characteristic of SO_2 pollution is that, at high humidities, it causes the stomata of certain plants to open wider than would normally occur in 'clean' air. This effect persists down to concentrations as low as 20 nl l^{-1} and means that at high humidities more SO_2 (and CO_2) enters leaf cells than normally would be the case. Mixtures of SO_2 and NO_2, however, close stomata at humidities at which SO_2 alone would cause opening. This finding in itself is sufficient to cause a major interactive effect on growth, etc.

By contrast, O_3 does not cause stomatal opening but may interact with other gases or stresses and accentuate their effects because the effect of this gas is primarily upon the outward-facing cell membranes. Reactions of O_3 with hydrocarbons, etc., then form intermediates which induce cells to leak water or ions (see Chapter 6). Consequently, problems of osmotic and ionic regulation arise which resemble very closely those caused by mixtures of SO_2 and NO_2. This probably means that small amounts of both sulphite and nitrite inside biological tissues are capable of producing free radicals (e.g. $HSO_3^{\cdot}$) where separately they would be unable to do so. These active species then act upon membranes, causing them to show leakiness exactly like the products from O_3 do. Furthermore, osmotic and ionic disturbances in a cell mean that less ATP, etc., is formed by chloroplasts and mitochondria. Consequently, extra energy is required to repair damage done by free radicals and again this is unavailable for growth, reproduction, etc. This energy diversion mechanism mediated by HHPs or free radicals is, so far, the best explanation of the more-than-additive effects of SO_2, NO_2 and O_3 upon plant growth.

Do 'cocktails' affect humans?

Generally, plants are more sensitive to atmospheric pollutants than animals although exceptions like H_2S and CO exist. Nevertheless, when the basis modes of action at the molecular level are examined, the similarities are greater than the differences. Very often, distinctions between them are only associated with specific functions such as photosynthesis in plants or local hormone and immunological responses in animals.

However, a wealth of detail exists from the study of plants to show that interactions between different atmospheric pollutants and stresses are important aspects of all air pollution effects and cannot be ignored. Rarely in studies of animals and health, however, is there an equivalent background. Little information exists on the possibility of interactions taking place in animals despite the fact that plants and animals (and people) inhabit the same polluted environment.

Explanations are easily provided. Studies of human health in the presence of pollutants are usually carried out by extensive epidemiological surveys which use medical records from critical localities. For more basic mechanistic studies, the use of experimental animals (with their attendant problems of extrapolation to the human condition) or human volunteers is difficult even with one pollutant and does not usually permit studies of mixtures of pollutants.

Other chapters have already made the point that when making accurate epidemiological surveys it is difficult to separate the effects of smoking, poverty, etc., from those of atmospheric pollution. Nevertheless, there is an awareness among medical statisticians that in some areas there are higher rates of infant mortality and bronchitis than one pollutant acting alone could be expected to cause. Most recognize that mixtures (or a 'cocktail' as they call it) of pollutants cause problems with health on a larger scale than was previously thought possible. In a sense, the CEC Limit Values (Table 1.7) are the first recognition of this possibility in tangible form because atmospheric levels of SO_2 are now regulated by the level of smoke that coexists with SO_2. If experience with interactions in plant science over the past two decades is also taken into account, international guidelines of air quality might well develop so that harmful coincidences of SO_2, smoke, acidic precipitation, nitrogen oxides, volatile hydrocarbons, and O_3 are identified and avoided.

Integration

Models

One of the most overworked words in the English language is the word model. Biologists, for example, study model systems (i.e. a life-cycle or ecosystem) to develop ideas and concepts while meteorologists model the possible implications of climate change. Some of these may be expressed mathematically or as a flow diagram using arrows to indicate inter-dependencies. These assist the appreciation of what is important, or what is trivial, and indicate what might be interesting to investigate next.

Understanding the different effects of atmospheric pollution on living systems presents environmental scientists with some complex problems. We know less than 10% of what concerns living systems and, similarly, less than 10% about all possible air pollution effects. This means that it is only on the basis of a small fraction that projections are made, implications assessed, and appropriate action advised. Nevertheless, this has to be done – it would be foolish to delay. Problems stemming from atmospheric pollution are with us now and some of them are of major global concern.

Vegetational impact

There are a number of models which attempt to understand the effects of air pollutants upon plants. Some of the best examples have been produced to illustrate the different hypotheses to explain recent forest decline. However, none of them attempts an overview where external changes to a plant are simultaneously considered in relation to possible internal responses. One attempt is shown in Fig. 11.1. The following paragraphs attempt to interpret this model (despite its apparent complexity) by using alphabetical identifiers for each stage on Fig. 11.1 and in the text.

Homogenic emissions of SO_2, nitrogen oxides, and hydrocarbons in bright sunlight cause the formation of secondary pollutants like O_3, etc. (A). In time, both primary and secondary pollutants are transferred to terrestrial and aquatic ecosystems by dry or wet deposition. Rates of dry deposition (B) vary widely but they are of similar magnitudes for all the major atmospheric pollutants. Meanwhile, a complex series of oxidations increase acidification (C) within the atmosphere. The amount of injury caused directly by wet deposition (D) on vegetation is still unresolved but surface oxidation within the liquid film continues. This may cause leaching of critical ions and modification of the rain falling through tree branches onto the soil (E). Acidic deposition also leaches calcium and magnesium from soils which mobilizes aluminium ions. These may influence root uptake of both nutrients and water (F).

Over 90% of the total pollutant flux during dry deposition enters by means of stomata (G) and the products pass into cells by a series of mechanisms collectively known as the mesophyll resistance (H). Effects of pollutants upon membranes (J) are still not understood but acidic gases and their products are known to have effects on the plasma membrane (K), the tonoplast (M) and the inner membranes (L), especially those of chloroplasts and mitochondria. Furthermore, SO_2 and nitrogen oxides raise levels of sulphite and nitrite inside cells which enhance free radical or HHP formation and leakiness very like O_3 does.

These products also affect the internal cell components (N), especially the cytoplasm (P), plastids (Q), and mitochondria (R). However, exchanges across the tonoplast (M) between the cytoplasm (P) and the vacuoles (O) during detoxification and pH readjustment are important as are exchanges between organelles and the cytoplasm for the cellular control of photosynthesis (Q) and respiration (R). Both processes contribute to the bioenergetic status of the cell and to the availability of energy (S) expended either in storage, metabolism or biosynthesis (T).

Detrimental effects on membranes (J–M) and cellular metabolism (N–T) are reflected by overall physiological changes (U). These include changes in extracellular events (W) associated with transport and translocation as well as nutrient uptake and xylem flow. The overall

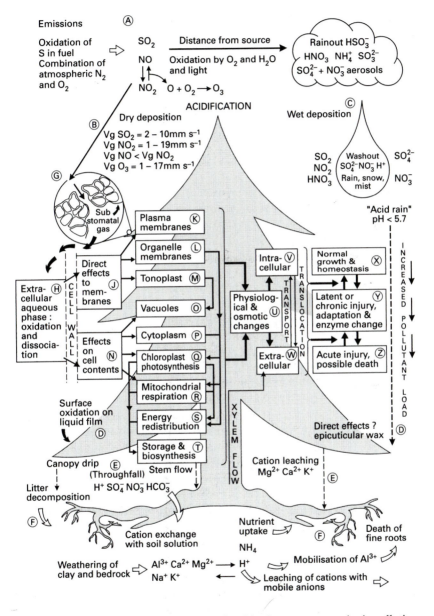

Fig. 11.1. Overview diagram showing relationships between atmospheric pollution and vegetation (see adjacent text for a full description; courtesy also of Dr P. H. Freer-Smith, UK Forestry Commission and CRC Press Inc., Florida).

success of the biochemical (K–T) readjustments, and the physiological controls that stem from them (U–W), is mirrored by the final disposition of the various cellular components.

With unaffected plants, this induces homeostatic readjustments (X) which then lead to normal growth, reproduction, and differentiation. Excess pollution, however, causes chronic damage (Y) or so-called 'invisible injury' which is a consequence of the diversion of energy from growth to repair. The effectiveness by which adaptation or tolerance is accomplished is measured by the balance between processes X and Y. Failure to achieve a balance leads to acute damage or 'visible injury' and possible death of the plant (Z).

Plants versus animals

At the cellular, biochemical or molecular level, it is most noticeable that there are great similarities in response of both plants and animals to atmospheric pollution. This is revealed if the overlying physiological differences between animals and plants are set aside. The passive gas uptake and circulatory systems of plants differ distinctly from active ventilation of the lungs (or gills) and circulation of the blood in animals. Yet, apart from these physiological differences, the mode of entry of pollutants into living cells and a wide range of the biochemical reactions thereafter are, by and large, very similar – especially those which detoxify free radicals. Indeed, study of the effects of air pollutants on plants has many implications for animal studies and *vice versa,* although this is rarely recognized by the two groups of researchers.

Basic differences between animals and plants in the uptake and evolution of O_2 and CO_2 are fundamental to life and, to a certain extent, overshadow all else when gaseous exchanges are considered. Both plants and animals contain haem compounds which, in most cases, are highly protected (e.g. cytochromes). However, haemoglobin in the blood of animals binds CO and H_2S more avidly than O_2 which accounts for the sensitivity of animals to these pollutants.

In the case of SO_2, nitrogen oxides and O_3, however, plants are more sensitive. Why should this be so? The best explanation is given by comparison of two calculations, both of which contain a number of approximations. If polluted air containing 100 nl l^{-1} O_3 is the starting point in each case, then it is possible to compare likely doses of a pollutant to both human lungs and plant mesophyll cells.

First we undertake the animal calculation. At normal temperatures and pressure, 22.4 l of polluted air contains 2.24 μl (or 0.1 μmol) of pollutant. If the average human ventilation rate at rest is 480 l h^{-1}, assuming a tidal volume of 500 ml and 16 ventilations min^{-1} (or 960 h^{-1}) then the intake of pollutant to the lungs over an hour would be 2.14 μmol of O_3. Under conditions of heavy exercise, the breathing rate may increase eight-fold (over the resting rate) and the expiratory reserve lung volume of 1.5 l is used in addition to the tidal volume of 500 ml. This gives a dramatic increase in ventilation rate (15 360 l h^{-1}) and intake of pollutant (69 μmol h^{-1}). Taking the alveolar surface area to be 100 m^2 and the thickness of the cell layer between the alveolar space and the blood to be 0.5 μm, then the volume of this layer would be 50 ml and the dosage to it therefore ranges from 2.18 μmol 50 ml^{-1} h^{-1} (or 43 μmol l^{-1} h^{-1}) at rest to 1.38 mmol l^{-1} h^{-1} during strenuous exercise.

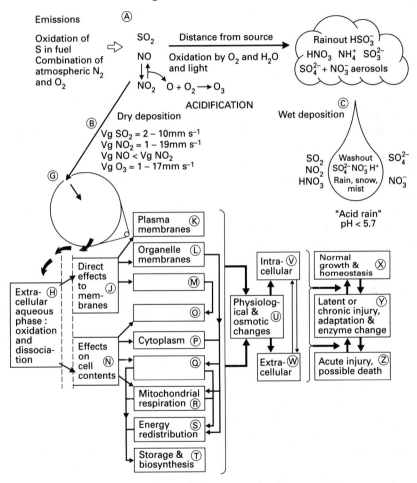

Fig. 11.2. The common core of Fig. 11.1 with plant-like features removed prior to conversion into Fig. 11.3.

Now we do the same for plants. In the case of O$_3$, uptake into plants largely depends upon stomatal access and the internal area of mesophyll cells available for uptake. This is between 5 and 30 times greater than the external surface area of a leaf. Assuming the boundary layer has been stripped away by sufficient wind (i.e. greater than 4.8 km h^{-1}) then, for a plant with a leaf surface of 14 m^2, the uptake rate through the stomata of 100 nl l^{-1} O$_3$-polluted air travelling with a speed of 4.8 km h^{-1} has been calculated to be 3.8 mmol h^{-1}. If the exposed internal cell surface area is taken as 25 times that of the leaf surface area and the thickness of the affected area is again 0.5 μm (i.e. a volume of 175 ml), it follows that the dosage to this sensitive area would be 3.8 mmol 175 ml^{-1} h^{-1} or 21.7 mmol l^{-1} h^{-1}.

Consequently, plant cells are exposed to more than 16 times as much pollutant as alveolar tissues of humans engaged in strenuous exercise even though the latter have active ventilation mechanisms. Differences of this magnitude explain rather better why plants are usually more sensitive than animals.

Humans, fishes, and other animals

One measure of the success of a paper model is that it must be flexible and, with the minimum of modification, capable of adapting to cover fresh situations. In the absence of a conceptual overview of changes in animals due to atmospheric pollution elsewhere in the literature, the plant-specific features of Fig. 11.1 (roots, photosynthesis, stomata, etc.) have been removed (leaving Fig. 11.2) and then animal-like features such as lungs, fish gills, immune and circulatory systems inserted (Fig. 11.3). The result gives a satisfactory working model (again using the same alphabetic identifiers on Fig. 11.3 and in the text).

In Fig. 11.3, fish gills show greatest response to acidic wet deposition (C) in run-off (D) and to interference by aluminium ions (E). On the other hand, the active inspiration (B) and expiration of breathing by land animals are more affected by dry deposition. Uptake into the alveolar cells from the alveolar spaces (G) is through an extracellular liquid film (H) and, once again, the effects are on membranes (J) or cell contents (N).

Free radical scavenging (Q) is evident in both animal and plant tissues, but effects on local modulators (M) may be important in any adjustments to physiological and osmotic changes (U) transmitted to the blood plasma (W, e.g. nitrogen oxides). Similarly, the effectiveness of immune responses (M), particularly those concerned with maintaining lung sterility, is known to be reduced in response to atmospheric pollutants such as SO$_2$.

The active circulation of the blood permits effects of pollutants upon the alveolar cells to be modulated by the activities of other body tissues. This may mean detoxification and elimination by the kidneys or, for example, reduced calcification of developing bone tissues in fish fry.

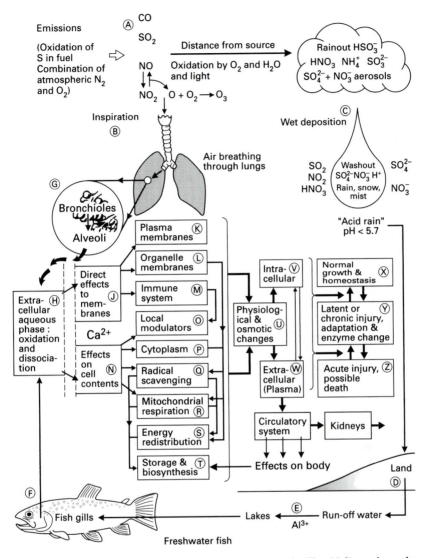

Fig. 11.3. Overview diagram derived from Fig. 11.1 (via Fig. 11.2) to show the relationships between atmospheric pollution and animals. Once again the text covers the individual details.

Despite such details, the essence of Fig. 11.3 remains the same as Fig. 11.1. Any interference to the physiological and osmotic changes (U) caused by the action of the pollutant products on processes H to T, either inside (V) or outside (W) the alveolar (or gill) cells, then causes normal growth and homeostasis (X) to be translated into chronic (Y) and acute (Z) injury.

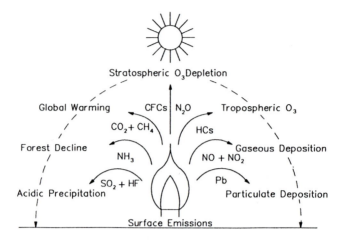

Fig. 11.4. Emissions of different pollutants may be involved in several direct and interactive effects upon the biosphere, some of which are understood and some still unsuspected.

Future implications

The possibility of interactions between different air pollutants and the various components of climate change cannot now be ignored. There is only one atmosphere and, as previous chapters have emphasized, there are a wide variety of processes going on in the atmosphere which have a multitude of effects on the biosphere. None of these are isolated from each other – all have the possibility of changing responses in each other (Fig. 11.4). A good example of this is shown by O_3. Depletion in the stratosphere increases the flux of UV-B, but tropospheric O_3 is rising in many places, which absorbs some of this UV-B, but in so doing adds to global warming.

How many more of these multiple interplays exist? It is clear that our present civilization has an increased tendency towards complexity, and when some of the consequences enter natural systems the outcome is far from simple or predictable. In the case of CFCs, for example, a whole series of fresh potential problems has emerged. Some of these particular implications we now understand but in the case of recent forest decline, for example, we still await a convincing elucidation of the cause(s). When we now consider possible implications it is clear we have to have a wider perspective and expectation of interactions as well as an understanding which enables us to delve further into the various mechanisms that cause the problems.

Further reading

Darrall, N.M. The effect of air pollutants on physiological processes in plants. *Plant, Cell and Environment* **12**, 1989, 1–30.

Koziol, M.J. and Whatley, F.R. (eds) *Gaseous Air Pollutants and Plant Metabolism.* Butterworths, 1984, London.

Lefohn, A.S. and Ormrod, D.P. *A Review and Assessment of the Effects of Pollutant Mixtures on Vegetation: Research Recommendations.* US Environmental Protection Agency, EPA600/3-84-037, 1984, Corvallis, Oregon.

National Institute for Environmental Studies. *Studies on Effects of Air Pollutant Mixtures on Plants.* Parts 1 and 2, NIESJ Nos. 65 and 66, 1984, Ibaraki, Japan.

Roberts, T.M., Darrall, N.M. and Lane, P. Effects of gaseous air pollutants on agriculture and forestry in the UK. *Advances in Applied Biology* **9**, 1983, 1–142.

Treshow, M. (ed.) *Air Pollution and Plant Life.* John Wiley & Sons, 1984, Chichester.

Appendix 1

Free radicals

'A radical is a man with both feet planted firmly in the air'
spoken by Franklin D. Roosevelt (broadcast 26 Oct. 1939).

The term 'free radical' or simply 'radical' is used to describe those chemical species capable of independent existence which have one or more unpaired electrons in their outer electronic orbitals. Chemical bonds are formed when electrons 'pair up' and this mainly explains why free radicals are highly reactive. Some radicals may also carry an electrical charge and, as a consequence, their properties are also influenced by charge.

There are several charged and neutral free radicals involved in air chemistry and atmospheric pollution. One example of a charged free radical is superoxide ($\cdot O_2^-$) – the negative superscript signifying the charge and the dot indicating that it is a free radical because it has an unpaired electron in an outer orbit around the nucleus. If one more electron is added to this outer orbital, then the free radical nature is lost and the molecule becomes peroxide (O_2^{2-}).

The most stable form of oxygen is 'ground-state' O_2. This itself is unusual in that it has one electron in each of its outer orbitals. By definition, ground-state O_2 is therefore a free radical in its own right and more reactive than $\cdot O_2^-$. Even more reactive forms of the oxygen molecule may also exist. These are often called 'singlet' oxygen and may or may not be free radicals.

Singlet oxygen is most readily formed during reaction of certain molecules with light (e.g. chlorophyll during photosynthesis) which then allows the transfer of this photochemical energy to a nearby oxygen molecule. The most frequent biological reaction with singlet oxygen is attack upon carbon–carbon double bonds to form peroxides.

Oxygen may be reduced in a number of different ways (see Fig. 6.7). A one-electron reduction gives $\cdot O_2^-$ (Reaction A1.1) and that with two electrons forms O_2^{2-} (Reaction A1.2). A four-electron reduction gives water (Reaction A1.3), but the splitting of H_2O_2 (the protonated form of O_2^{2-}) in the light (Reaction A1.4) or in the presence of iron or copper (Reaction A1.5) forms one of the most reactive radicals known, that is, $\cdot OH$.

$$O_2 + e \Rightarrow {}^{\bullet}O_2{}^{-} \tag{A1.1}$$

$$O_2 + 2e \Rightarrow O_2{}^{2-} \tag{A1.2}$$

$$O_2 + 4H^+ + 4e \Rightarrow 2H_2O \tag{A1.3}$$

$$H_2O_2 + light \Rightarrow 2{}^{\bullet}OH \tag{A1.4}$$

$$Cu^+ \text{ or } Fe^{2+} + H_2O_2 \Rightarrow Cu^{2+} \text{ or } Fe^{3+} + {}^{\bullet}OH + OH^- \tag{A1.5}$$

Monoatomic oxygen, like some other atomic forms of gases (hydrogen, fluorine, nitrogen atoms), is also a free radical although the dot superscript is rarely used to indicate the presence of unpaired electrons – it is merely assumed. It too is capable of reacting with water to form ${}^{\bullet}OH$ radicals (Reaction A1.6).

Once formed, a free radical reacts with other components (e.g. Reaction A1.6 involving H_2O) to form other free radicals which may be more or less reactive than the original radical. An example of the former is reaction of ${}^{\bullet}OH$ with H_2O_2 (Reaction A1.7) to form less reactive ${}^{\bullet}O_2{}^{-}$. However, reaction with anions such as fluoride (Reaction A1.8) forms fluorine atoms or more reactive fluoride radicals (Reaction A1.9) which are as reactive as ${}^{\bullet}OH$ radicals and react with virtually any organic or inorganic component of biological systems.

$$O + H_2O \Rightarrow 2{}^{\bullet}OH \tag{A1.6}$$

$${}^{\bullet}OH + H_2O_2 \Rightarrow H_2O + H^+ + {}^{\bullet}O_2{}^{-} \tag{A1.7}$$

$$F^- + {}^{\bullet}OH \Rightarrow F^{\bullet} + OH^- \tag{A1.8}$$

$$F^{\bullet} + F^- \Rightarrow F_2^{\bullet -} \tag{A1.9}$$

Appendix 2

Acidity, pH, pK_a and microequivalents

'Razors pain you; rivers are damp; acids stain you; and drugs cause cramp.
Guns aren't lawful; nooses give; gas smells awful; you might as well live.'
from 'Résumé' by Dorothy Parker (1893–1967).

A substance that is able to donate H^+ ions or protons (more strictly, hydroxonium ions, H_3O^+) to a solution is defined as an acid. Conversely, a base is a substance that accepts H^+ ions. These definitions are useful because they parallel those used to define reduction and oxidation and emphasize the close thermodynamic relationships that exist between the two phenomena.

By convention, a substance is enclosed within square brackets if a concentration of that substance is also implied: thus $[H^+]$ indicates a concentration of H^+ ions in moles per litre. In Reaction A2.1, for example, the equilibrium constant ($K_{A2.1}$) governing the ionization of acetic acid is given by the following relationship:

$$CH_3COOH \Leftrightarrow CH_3COO^- + H^+ \qquad \text{(A2.1)}$$

$$K_{A2.1} = \frac{[CH_3COO^-][H^+]}{[CH_3COOH]}$$

Therefore, if the $[H^+]$ and the equilibrium constant are known (i.e. $K_{A2.1} = 2.24 \times 10^{-5}$ M), it is possible to calculate the ratio between the charged and uncharged forms of acetic acid. Another way of looking at this is to appreciate that the concentration of the charged and uncharged forms is governed by the $[H^+]$.

The usual way of expressing $[H^+]$ is on a logarithmic scale, known as the pH scale, because very large ranges of concentrations may exist. In this case the pH is defined as

$$pH = \log\left(\frac{1}{[H^+]}\right) = -\log[H^+]$$

This gives a scale of pH values from 0 to 14 but biological systems usually involve pH changes within the range 2 to 9. If, in the acetic acid

Table A2.1 pK_a values for acids in aqueous solution

System	Equilibrium		pK_a
Sulphuric acid	H_2SO_4	$\Leftrightarrow H^+ + HSO_4^-$	< -2
Nitric acid	HNO_3	$\Leftrightarrow H^+ + NO_3^-$	-1.4
Sulphurous acid	H_2SO_3	$\Leftrightarrow H^+ + HSO_4^-$	1.8
Bisulphate	HSO_4^-	$\Leftrightarrow H^+ + SO_4^{2-}$	2.0
Hydrofluoric acid	HF	$\Leftrightarrow H^+ + F^-$	3.3
Nitrous acid	HNO_2	$\Leftrightarrow H^+ + NO_2^-$	3.3
Aluminium	$Al(H_2O)_6^{3+}$	$\Leftrightarrow H^+ + Al(H_2O)_5(OH)^{2+}$	5.0
Carbonic acid	$H_2O + CO_2$	$\Leftrightarrow H^+ + HCO_3^-$	6.4
Hydrogen sulphide	H_2S	$\Leftrightarrow H^+ + HS^-$	7.1
Bisulphite	HSO_3^-	$\Leftrightarrow H^+ + SO_3^{2-}$	7.2
Ammonium	NH_4^+	$\Leftrightarrow H^+ + NH_3$	9.3
Bicarbonate	HCO_3^-	$\Leftrightarrow H^+ + CO_3^{2-}$	10.3
Bisulphide	HS^-	$\Leftrightarrow H^+ + S^{2-}$	12.9
Water	H_2O	$\Leftrightarrow H^+ + OH^-$	14.0

example, the charged $[CH_3COO^-]$ and uncharged $[CH_3COOH]$ forms of the acid are equal to each other it then follows that

$$[H^+] = K_{A2.1}$$

The pH at which this occurs is known as the pK_a value and is the point of maximum sensitivity to pH change. In this case, as $K_{A2.1}$ is equivalent to 2.24×10^{-5} M, then this particular pH or pK_a value is 4.65. The pK_a value is also a useful indicator of the strength of an acid. The stronger the acid the more ready it is to donate H^+ ions and hence more H^+ ions are required to reverse this action. A number of important ionizations relevant to air pollution and their pK_a values are shown in Table A2.1.

Values of pK_a in relation to pH are useful as a quick guide because a move of one pH unit away from the pK_a value changes the ratio between the charged and uncharged forms of an acid by a factor of 10. One way of calculating these changes more accurately over the whole range of pH values, especially if they do not differ by whole pH numbers, is to use the Henderson–Hasselbalch equation which has a number of different forms summarized as follows:

$$pH = pK_a + \log\left(\frac{[salt]}{[acid]}\right)$$

or

$$pH = pK_b + \log\left(\frac{[base]}{[salt]}\right)$$

In the case of Reaction A2.1, if the pH is measured as 2.65, the following equation is relevant:

$$pH = pK_a + \log\left(\frac{[CH_3COO^-]}{[CH_3COOH]}\right)$$

Substituting gives

$$2.65 = 4.65 + \log\left(\frac{[CH_3COO^-]}{[CH_3COOH]}\right)$$

or

$$-2 = \log\left(\frac{[CH_3COO^-]}{[CH_3COOH]}\right)$$

Therefore:

$$antilog(-2) = \frac{[CH_3COO^-]}{[CH_3COOH]} = 0.01$$

This means that the ratio of uncharged to charged acid is $1:100$. On back checking, this must be so because an increase of $[H^+]$ 100-fold over the value at the pK_a has taken place.

Ionization of H_2O (Reaction A2.2) is governed by the appropriate equilibrium constant, $K_{A2.2}$. But because $[H_2O]$ in relation to $[H^+]$ or $[OH^-]$ declines imperceptibly, it is often treated as a constant.[1] By multiplying $K_{A2.2}$ by $[H_2O]$ one then obtains the ionic product of H_2O (K_w) as follows:

$$H_2O \Leftrightarrow H^+ + OH^- \tag{A2.2}$$

$$K_{A2.2} = \frac{[H^+][OH^-]}{[H_2O]}$$

or

$$K_{A2.2} \times [H_2O] = K_w$$

---

[1] Conventionally, when equilibria involve H_2O as a reactant, $[H_2O]$ is actually set at 1 M and not 55.5 M (i.e. 1000 g l^{-1} divided by the molecular weight of 18.016). All equilibrium constants involving H_2O as a reactant therefore need to be adjusted by a factor of 55.56 (M) if other factors such as rate constants, etc., are under consideration.

or

$$K_w = [H^+][OH^-] = 10^{-14} \; M^2$$

So when $[H^+]$ equals that of $[OH^-]$ (i.e. $[H^+]^2$) this will give a neutral solution. Consequently, substituting in the above:

$$[H^+]^2 = 10^{-14} \; M^2$$

gives

$$[H^+] = 10^{-7} \; M$$

or

$$pH = 7$$

Therefore, at pH 0, the $[OH^-]$ will be negligible and $[H^+]$ will be 1 M for simple equilibria like those shown in Reaction A2.1.

Another way of expressing $[H^+]$ is in terms of equivalents because, in some cases, certain acidic reactions (e.g. Reaction A2.3) produce more H^+ ions than simpler reactions (e.g. Reactions A2.1 or A2.2) and the identity of the acid or its products is sometimes unknown. The relationship between pH and equivalents is then given by

$$H_2SO_4 \Leftrightarrow 2H^+ + SO_4^{2-} \tag{A2.3}$$

$$pH = 6.0 - \log \text{(}\mu\text{equivalents of } H^+ \; l^{-1}\text{)}$$

Therefore, at pH 0, there is one equivalent of H^+ per litre, at pH 3 one milliequivalent, and at pH 6 one µequivalent.

Appendix 3

Glossary and units

'The Aire[1] below is double dyed and damned;
the air above, with lurid smoke is crammed;
the one flows steaming foul as Charon's Styx[1]
its poisonous vapours in the other mix.'
These sable twins the murky town invest;
by them the skin's begrimed, the lungs oppressed.
How dear the penalty thus paid for wealth
obtained through wasted life and broken health.'
Quoted by William Osburn at a meeting of The Leeds Philosophical and Literary Society in 1857.

ATP	adenosine triphosphate, which releases large amounts of energy when it hydrolyses
BAF	biological amplification factor, the percentage increase in an effect (e.g. skin cancer) that results from each increment in annual UV dose
BaU	Business-as-Usual scenario: no regulatory action is taken
CAM	Crassulacean acid metabolism
CEC	Commission of the European Communities
CFCs	chlorofluorocarbons
–CHO	an aldehyde group
$>$C=O	a ketone group
DNA	deoxyribonucleic acid, the code of life
EPA	United States Environmental Protection Agency
HCs	hydrocarbons
HCFCs	hydrochlorofluorocarbons, having less chlorine or fluorine than CFCs
HFAs	hydrofluoroalkyls, replacements for CFCs and HCFCs
HHPs	hydroxyhydroperoxides, products derived from ozonolysis of HCs, etc.
IAA	indoleacetic acid, a plant hormone involved in cell proliferation
IPCC	UN Intergovernmental Panel on Climate Change
IR	infrared radiation starting from 700 nm upwards (see Table 7.1)
LD_{50}	lethal dose which kills 50% of the population tested
N	general abbreviation for all nitrogen-containing compounds involved

[1] Rivers flowing through Leeds, and nine times round the Underworld of Greek mythology, respectively.

NASA	United States National Aeronautics and Space Administration
OECD	Organization for Economic Cooperation and Development
PEP	phospho-enol-pyruvate, a C_3 carboxylic acid
ppm, ppb	non-SI units of concentration; when used in the context of atmospheric pollution they are now superseded by the use of $\mu l\ l^{-1}$ and $nl\ l^{-1}$ respectively (see Tables A3.1 and A3.2)
RAFs	radiation amplication factors, measures of the biological significance of a solar radiation change in a particular range (e.g. 280–315 nm) brought about by a specified % atmospheric O_3 depletion
RBCs	red blood cells or corpuscles, also known as erythrocytes
RNA	ribonucleic acid, the intermediary between DNA and proteins
S	general abbreviation for all sulphur-containing compounds involved
SI	Système International (d'Unités) – see note on Units
SOD	superoxide dismutase, enzyme which breaks down $^{\bullet}O_2^{-}$ radicals
TLVs	threshold limit values, which are one type of occupational exposure limit (OEL) derived by the American Conference of Government Industrial Hygienists (ACGIH). In fact, there are several types of TLVs. The types quoted here are TLV-TWAs (time-weighted averages) for 8 h work days in a 40 h week, but TLV-STELs (short-term exposure limits) are also used, along with permitted exposure limits (PELs) derived by the US Occupational Safety and Health Administration (OSHA), and recommended exposure limits (RELs) from the US National Institute for Occupational Safety and Health (NIOSH). Most other countries use TLV-TWAs for air pollution control programmes
UNEP	United Nations Environment Programme
UV-A	ultraviolet radiation (UV) from 315 to 400 nm (see Table 7.1)
UV-B	ultraviolet radiation (UV) from 280 to 315 nm
UV-B_{BE}	biologically effective UV-B
WHO	World Health Organization
WMO	World Meteorological Organization

Units

The units used throughout this book conform to the metric 'Système International (SI) d'Unités', the base units of which are:

m	metre	K	kelvin (temperature)
kg	kilogram	mol	mole (amount of substance)
s	second	cd	candela (luminosity)
A	ampere (current)		

Some of the multiples of SI are familiar to most, e.g.

k	kilo (10^3)	m	milli (10^{-3})
M	mega (10^6)	μ	micro (10^{-6})

but others are less well known, e.g.

G	giga (10^9)	n	nano (10^{-9})
T	tera (10^{12})	p	pico (10^{-12})
P	peta (10^{15})	f	femto (10^{-15})
E	exa (10^{18})		

Other units used in this book are derived SI units, and include:

Bq   becquerel (number of radioactive disintegrations per unit time)
ha   hectare (10^4 m^2)
°C   degrees Celsius (0°C = 273.15 K)
J   joule (unit of heat, energy or work $\equiv$ 1 kg m^{-1} s^{-2} [the newton, N] m^{-1})
l   litre (0.001 m^3)
Sv   sievert (dose equivalent of radioactivity)
t   metric tonne (1000 kg)
V   volt (electrical potential difference $\equiv$ 1 J A^{-1} s^{-1})
W   watt (power $\equiv$ 1 J s^{-1})

The only non-SI units used knowingly are:

a   annum (year)
h   hour
d   day

Table A3.1 Amounts of different pollutants in $\mu g\ m^{-3}$ equivalent to $1\ \mu l\ l^{-1}$ ($1000\ nl\ l^{-1}$) at different temperatures but standard atmospheric pressure

Temperature (°C)	SO_2	NO_2	NO	NH_3	O_3	CS_2	CO	H_2S	HF
−5	2914	2092	1365	774	2183	2001	1274	1550	910
0	2860	2054	1340	760	2143	1965	1250	1521	893
5	2809	2017	1316	747	2104	1929	1228	1494	877
10	2759	1981	1292	733	2067	1895	1206	1468	862
15	2711	1947	1270	721	2031	1862	1185	1442	847
20	2665	1914	1248	708	1997	1831	1165	1417	832
25	2620	1882	1227	696	1963	1800	1146	1394	818
30	2577	1850	1207	685	1931	1770	1127	1371	805
35	2535	1821	1187	674	1899	1742	1108	1348	792

Table A3.2 Amounts of different pollutants in $nl\ l^{-1}$ equivalent to $1000\ \mu g\ m^{-3}$ at different temperatures but standard atmospheric pressure

Temperature (°C)	SO_2	NO_2	NO	NH_3	O_3	CS_2	CO	H_2S	HF
−5	343	478	733	1291	458	500	785	645	1099
0	349	487	746	1315	467	509	800	657	1120
5	356	496	760	1339	475	518	814	669	1140
10	362	505	774	1364	484	528	829	681	1161
15	368	514	787	1388	492	537	844	693	1181
20	375	523	801	1412	501	546	858	705	1202
25	382	531	815	1436	509	556	873	718	1222
30	388	540	828	1460	518	565	888	730	1243
35	394	549	842	1484	527	574	902	742	1263

INDEX

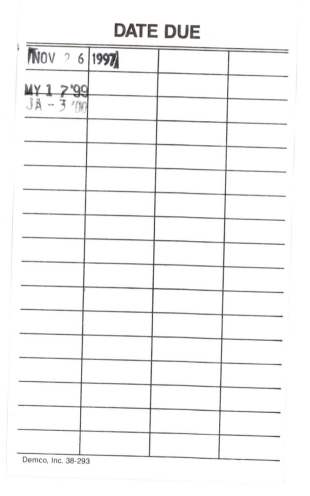

DATE DUE

NOV 2 6 1997			
MY 1 7 '99			
JA – 3 '00			